Ceramic Materials for
Energy Applications II

Ceramic Materials for Energy Applications II

A Collection of Papers Presented at the 36th International Conference on Advanced Ceramics and Composites
January 22–27, 2012
Daytona Beach, Florida

Edited by
Kevin Fox
Yutai Katoh
Hua-Tay Lin
Ilias Belharouak

Volume Editors
Michael Halbig
Sanjay Mathur

The American Ceramic Society

A John Wiley & Sons, Inc., Publication

Published by John Wiley & Sons, Inc., Hoboken, New Jersey.
Published simultaneously in Canada.

For general information on our other products and services or for technical support, please contact our Customer Care Department within the United States at (800) 762-2974, outside the United States at (317) 572-3993 or fax (317) 572-4002.

Wiley also publishes its books in a variety of electronic formats. Some content that appears in print may not be available in electronic formats. For more information about Wiley products, visit our web site at www.wiley.com.

Library of Congress Cataloging-in-Publication Data is available.

ISBN: 978-1-118-20599-0
ISSN: 0196-6219

Printed in the United States of America.

10 9 8 7 6 5 4 3 2 1

Contents

Preface   vii

Introduction   ix

Analytical Techniques for Li-S Batteries   1
  Manu U. M. Patel, Rezan Demir Cakan, Mathieu Morcrette,
  Jean-Marie Tarascon, Miran Gaberscek, and Robert Dominko

Three New Approaches using Silicon in Three Valuable Energy   11
Applications
  John Carberry

Processing of Inert SiC Matrix with TRISO Coated Fuel by Liquid   25
Phase Sintering
  Kazuya Shimoda, Tatsuya Hinoki, Kurt A. Terrani, Lance L. Snead,
  and Yutai Katoh

SiC-Coated HTR Fuel Particle Performance   33
  Michael J. Kania, Heinz Nabielek, and Karl Verfondern

Study of the Silicon Carbide Matrix Elaboration by Film Boiling   71
Process
  Aurélie Serre, Joëlle Blein, Yannick Pierre, Patrick David, Fabienne Audubert,
  Sylvie Bonnamy, and Eric Bruneton

Processing of Ultrafine Beta-Silicon Carbide Powder by   85
Silicon–Carbon Reaction
  S. Sonak, S. Ramanathan, and A. K. Suri

Characterization of Failure Behavior of Silicon Carbide Composites   95
by Acoustic Emission
  Takashi Nozawa, Kazumi Ozawa, and Hiroyasu Tanigawa

Recession of Silicon Carbide in Steam under Nuclear Plant LOCA   111
Conditions up to 1400 °C
    Greg Markham, Rodney Hall, and Herbert Feinroth

The Effect of Temperature and Uniaxial Pressure on the   121
Densification Behavior of Silica Aerogel Granules
    J. Matyáš, M. J. Robinson, and G. E. Fryxell

Microstructural Analysis of Nuclear Grade Graphite Materials   133
    Kentaro Takizawa, Toshiaki Fukuda, Akira Kondo, Yutai Katoh,
    and G. E. Jellison

A Model for Simulation of Coupled Microstructural and   145
Compositional Evolution
    Veena Tikare, Eric R. Homer, and Elizabeth A. Holm

Characterisation of Corrosion of Nuclear Metal Wastes Encapsulated   159
in Magnesium Silicate Hydrate (MSH) Cement
    Tingting Zhang, Chris Cheeseman, and Luc J. Vandeperre

Impact of Uranium and Thorium on High TiO_2 Concentration   169
Nuclear Waste Glasses
    Kevin M. Fox and Thomas B. Edwards

Author Index   181

Preface

This proceedings issue contains contributions from three energy related symposia and the European Union–USA Engineering Ceramics Summit that were part of the 36th International Conference on Advanced Ceramics and Composites (ICACC), in Daytona Beach, Florida, January 22-27, 2012. The symposia include Ceramics for Electric Energy Generation, Storage and Distribution; Advanced Ceramics and Composites for Nuclear and Fusion Applications; and Advanced Materials and Technologies for Rechargeable Batteries. These symposia and the Summit were sponsored by the ACerS Engineering Ceramics Division. The symposium on Advanced Ceramics and Composites for Nuclear and Fusion Applications was cosponsored by the ACerS Nuclear & Environmental Technology Division.

The editors wish to thank the authors and presenters for their contributions, the symposium organizers for their time and labor, and all the manuscript reviewers for their valuable comments and suggestions. Acknowledgment is also due for financial support from the Engineering Ceramics Division, the Nuclear & Environmental Technology Division, and The American Ceramic Society. The editors wish to thank Greg Geiger at ACerS for all his effort in assembling and publishing the proceedings.

KEVIN FOX, Savannah River National Laboratory
YUTAI KATOH, Oak Ridge National Laboratory
HUA-TAY LIN, Oak Ridge National Laboratory
ILIAS BELHAROUAK, Argonne National Laboratory

Introduction

This issue of the Ceramic Engineering and Science Proceedings (CESP) is one of nine issues that has been published based on content presented during the 36th International Conference on Advanced Ceramics and Composites (ICACC), held January 22–27, 2012 in Daytona Beach, Florida. ICACC is the most prominent international meeting in the area of advanced structural, functional, and nanoscopic ceramics, composites, and other emerging ceramic materials and technologies. This prestigious conference has been organized by The American Ceramic Society's (ACerS) Engineering Ceramics Division (ECD) since 1977.

The 36th ICACC hosted more than 1,000 attendees from 38 countries and had over 780 presentations. The topics ranged from ceramic nanomaterials to structural reliability of ceramic components which demonstrated the linkage between materials science developments at the atomic level and macro level structural applications. Papers addressed material, model, and component development and investigated the interrelations between the processing, properties, and microstructure of ceramic materials.

The conference was organized into the following symposia and focused sessions:

Symposium 1	Mechanical Behavior and Performance of Ceramics and Composites
Symposium 2	Advanced Ceramic Coatings for Structural, Environmental, and Functional Applications
Symposium 3	9th International Symposium on Solid Oxide Fuel Cells (SOFC): Materials, Science, and Technology
Symposium 4	Armor Ceramics
Symposium 5	Next Generation Bioceramics

Symposium 6	International Symposium on Ceramics for Electric Energy Generation, Storage, and Distribution
Symposium 7	6th International Symposium on Nanostructured Materials and Nanocomposites: Development and Applications
Symposium 8	6th International Symposium on Advanced Processing & Manufacturing Technologies (APMT) for Structural & Multifunctional Materials and Systems
Symposium 9	Porous Ceramics: Novel Developments and Applications
Symposium 10	Thermal Management Materials and Technologies
Symposium 11	Nanomaterials for Sensing Applications: From Fundamentals to Device Integration
Symposium 12	Materials for Extreme Environments: Ultrahigh Temperature Ceramics (UHTCs) and Nanolaminated Ternary Carbides and Nitrides (MAX Phases)
Symposium 13	Advanced Ceramics and Composites for Nuclear Applications
Symposium 14	Advanced Materials and Technologies for Rechargeable Batteries
Focused Session 1	Geopolymers, Inorganic Polymers, Hybrid Organic-Inorganic Polymer Materials
Focused Session 2	Computational Design, Modeling, Simulation and Characterization of Ceramics and Composites
Focused Session 3	Next Generation Technologies for Innovative Surface Coatings
Focused Session 4	Advanced (Ceramic) Materials and Processing for Photonics and Energy
Special Session	European Union – USA Engineering Ceramics Summit
Special Session	Global Young Investigators Forum

The proceedings papers from this conference will appear in nine issues of the 2012 Ceramic Engineering & Science Proceedings (CESP); Volume 33, Issues 2-10, 2012 as listed below.

- Mechanical Properties and Performance of Engineering Ceramics and Composites VII, CESP Volume 33, Issue 2 (includes papers from Symposium 1)
- Advanced Ceramic Coatings and Materials for Extreme Environments II, CESP Volume 33, Issue 3 (includes papers from Symposia 2 and 12 and Focused Session 3)
- Advances in Solid Oxide Fuel Cells VIII, CESP Volume 33, Issue 4 (includes papers from Symposium 3)
- Advances in Ceramic Armor VIII, CESP Volume 33, Issue 5 (includes papers from Symposium 4)

- Advances in Bioceramics and Porous Ceramics V, CESP Volume 33, Issue 6 (includes papers from Symposia 5 and 9)
- Nanostructured Materials and Nanotechnology VI, CESP Volume 33, Issue 7 (includes papers from Symposium 7)
- Advanced Processing and Manufacturing Technologies for Structural and Multifunctional Materials VI, CESP Volume 33, Issue 8 (includes papers from Symposium 8)
- Ceramic Materials for Energy Applications II, CESP Volume 33, Issue 9 (includes papers from Symposia 6, 13, and 14)
- Developments in Strategic Materials and Computational Design III, CESP Volume 33, Issue 10 (includes papers from Symposium 10 and from Focused Sessions 1, 2, and 4)

The organization of the Daytona Beach meeting and the publication of these proceedings were possible thanks to the professional staff of ACerS and the tireless dedication of many ECD members. We would especially like to express our sincere thanks to the symposia organizers, session chairs, presenters and conference attendees, for their efforts and enthusiastic participation in the vibrant and cutting-edge conference.

ACerS and the ECD invite you to attend the 37th International Conference on Advanced Ceramics and Composites (http://www.ceramics.org/daytona2013) January 27 to February 1, 2013 in Daytona Beach, Florida.

MICHAEL HALBIG AND SANJAY MATHUR
Volume Editors
July 2012

ANALYTICAL TECHNIQUES FOR Li-S BATTERIES

Manu U.M. Patel, Rezan DEMIR CAKAN, Mathieu MORCRETTE, Jean-Marie TARASCON, Miran GABERSCEK, Robert DOMINKO

National Institute of Chemistry, Hajdrihova 19, SI-1000 Ljubljana, Slovenia
ALISTORE-ERI, 33 Rue Saint-Leu, 80039 Amiens, France
LRCS, Université de Picardie Jules Verne, 33 Rue Saint-Leu, 80039 Amiens, France

ABSTRACT
        Lithium sulfur rechargeable batteries are foreseen to be used in electric vehicles in the near future. To enable their placement to the market we need to improve their reliability and cycling performance. This can be done by selective change of the chemical environment in the Li-S battery, including electrolyte (solvents and salts), different combinations of cathode composites and different additives. With an aim to understand differences between different chemical environments, suitable and reliable analytical techniques that can effectively monitor the changes of interest, should be developed. In this work we present two newly developed analytical techniques which are capable to detect quantitative and qualitative differences between different polysulfide species in the electrolyte. More specifically, both proposed techniques can detect the polysulfides that have diffused away from the cathode composite. With a modified 4-electrode Swagelok cell we can quantitatively determine the amount of such polysulfides, while an in-situ UV-Vis cell gives us information about the composition of these polysulfides. Combining these techniques with a classical galvanostatic cycling method could lead to a better understanding, consequently, a faster tuning of Li-S battery properties.

INTRODUCTION
        The on-going and foreseen increased electrification of the transport sector, the "electromobility" revolution is one of the major driving forces for energy storage breakthroughs. Current rechargeable Li-ion batteries for electric vehicle (EV) are capable to deliver between 100 Wh/kg and 150 Wh/kg energy density, while typical consumption of a liter of gasoline produces 2500 Wh of useful work. So there is still a factor of 15 between the energy delivered by one liter of gasoline and 1 kg of battery (e.g. the autonomy of the car with similar weight that is driven by batteries is between 5-10 times shorter than with gasoline). Hence, if we want to achieve or even approach the goal of a 500 km driving range using battery powered vehicles, we need to explore new batteries that are different from the existing Li-ion technology and that offer a real step further in the energy storage[1].
        One of the possibilities is the lithium-sulfur battery technology, the principle of which has been known for several decades[2-4], however without real commercial breakthrough. In theory, Li-S battery can fulfill all the requirements of the intelligent vehicle battery system since it possesses a high volumetric (small size) and a high gravimetric (low weight) energy density. In addition, it can be produced as a flexible, environmental friendly and cost effective cell and in theory offers a safe and a reliable operation.
        Elemental sulfur as a positive electrode material in combination with lithium metal as a negative electrode material offers an attractive high-energy rechargeable cell. More specifically, assuming the whole reaction to Li_2S, the average redx voltage and the theoretical energy values of such a cell are 2.1V and 2500W/kg (or 2800Wh/l), respectively.
        One of the main reasons that the Li-S system has still not reached a wide commercial availability is the not-yet-optimized function of the cathode. Over the last two decades different directions towards improved cathode architecture and chemical compositions have been proposed. One of the strategies has been a special cell configuration where all polysulfides are solubilized – the so called catholyte cells[2]. Another strategy is the use of either a mixture of sulfur and suitable matrix to

1

improve the electronic conductivity or embedment of sulfur into a polymer matrix (for instance polyaniline).

The first discharge of Li-S battery and potential reactions are shown in the Figure 1. Along the high voltage plateau (2,3 – 2,4V), reduction of cyclic sulfur to the long chain polysulfides (e.g. Li_2S_8 and Li_2S_6) occurs. Further reduction progressively leads to the low voltage plateau at 2,1V and to the formation of Li_2S_4. The plateau at 2.1V corresponds to a 1 electron reaction per sulfur atom and to the reduction to Li_2S_2. Reduction of lithium disulfide to lithium monosulfide occurs in region C (Figure 1) and leads to a very fast voltage decay due to insolubility of Li_2S in most of electrolytes (due to the formation of insoluble blocking layer of Li_2S).

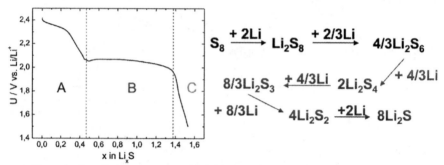

Figure 1: First discharge curve for Li-S battery and suggested reaction scheme divided into three typical regions of sulfur reduction

Regarding the chemical environment, most of polysulfides (except Li_2S) are soluble in electrolytes and can diffuse away from the cathode composite, thus leading to active material loss and self-discharge. Once dissolved in the electrolyte, they can move to the negative electrode (metallic lithium) and react with it, leading to a shuttle mechanism that decreases charge efficiency and leads to the formation of blocking Li_2S layer on the surface of lithium. With an aim to minimize the diffusion of soluble polysulfides from the cathode composite, different strategies have been proposed. The most promising direction of cathode composite architecture is the use of mesoporous matrix with a tiny layer of sulfur dispersed on the surface of pores. Pores in this case can act as mini chambers and at least in the formation cycles the diffusion of polysulfides is reduced and postponed to later stages of Li-S battery cycling. Nevertheless, mesoporous substrate represents a high surface area substrate for sulfur impregnation. The other possible solutions how to prevent diffusion of reduced states of sulfur are the use of different electrolyte combinations or the use of different additives in the electrolyte. The common point in all of the literature reports on that matter is the lack of detailed understanding about how different changes in the chemical environment (changes in the composition and architecture of cathode composite, changes in the electrolyte formulation, etc.) affect the electrochemical performance (utilization of sulfur and discharge/charge efficiency).

Whereas most of the published works focus predminantly on the cycling behavior, efficient progress is only possible if one deeply understands the correlation between the morphological and compositional changes on one side and the electrochemistry on the other. In our recent work[6], we have proposed the use of a modified 4-electrode Swagelok cell that can be used as a reliable analytical cell for the quantitative determination of polysulfides that have diffused away from cathode composite.

EXPERIMENTAL

Cathode composites (carbon + sulfur in 1:1 weight ratio) were prepared by infiltration of sulfur onto the carbon black (Printex XE2) surface or into the mesoporous carbon. Cathode composites were mixed with 7 wt.% of PVdF (Aldrich) and additional 8 wt.% of Printex XE2 carbon black and casted on an aluminum current collector. Electrodes with diameter 12 mm containing 2-3 mg of sulfur were pressed and dried at 90°C overnight before use.

Electrolyte solutions used in this study were 1M LiTFSI (lithium bis(trifluirimethanesulfonyl)imide) containing tetra methylene sulphone (TMS) or ethyl methyl sulphone (EMS). Electrolyte solutions with chemically synthesized polysulfides were prepared by mixing adequate amount of polysulfide powder with electrolyte solutions. Polysulfide powder was prepared by reaction of stoichiometric amounts of sulfur and lithium in ethylene glycol diethyl ether at 150°C. The denotation of polysulfides is based on the stoichiometric composition and it is not necessary to reflect exact electronic state of sulfur. Chemically synthesized polysulfides, which were used for standardization, were dissolved in electrolyte to form known concentration in the electrolyte. 1mM, 5mM, 10mM, 25mM and 50mM solutions were prepared by dissolving following Li_2S_n polysulfides (n=2,...,8).

The principle of 4 electrode Swagelok cell operation was published in our previous work[6]. The cell compromises two electrochemical cells in one, where one serves as a typical battery cell using two electrode cells (a working electrode with casted cathode composite and lithium as counter and reference electrodes). The second one is built perpendicularly to the battery sandwich and placed in between two separators, where nickel or stainless steel wires can be used as a working electrode and platinum wire as a counter and as a reference electrode. Operation of two cells is in the sequence, i.e. galvanostatic discharging with a C/20 rate for 2h of Li-S battery followed by a cyclovoltametric measurement using perpendicular electrodes in the potential region between 2.5 and 1 V with a scan rate of 2 mVs^{-1} and continued with galavanostatic discharging. Measurements with different concentration of Li_2S_8 were performed in the same way as in ordinary Li-S battery cells, except that the cathode composite did not contain any sulfur.

The cell designed for the in-situ UV-Vis measurements is based on the coffee bag cell with a sealed quartz window on one side. The cell was mounted in the UV-Vis spectrophotometer and measured simultaneously in the reflection mode during galvanostatic cycling. Charge/discharge rate was C/20 and UV-Vis spectra were measured every 15 min. Battery was built by using pressed electrode composite on the aluminum substrate, separator and ring based lithium electrode as a counter electrode. Design of ring based electrode is necessary for the detection of polysufides in the separator in the reflection mode. Electrolyte solutions with a known concentration of different polysulfides – Li_2S_n (n=2-8) were used for the calibration. UV-Vis spectra were measured in the cell designed for the electrochemical measurements (cell with a quartz window), except that only a separator with a wiped 1 mL of electrolyte solution containing different polysulfides was used.

RESULTS AND DISCUSSION

The applicability of modified 4-electrode Swagelok cell was tested first in the configuration of a blank cell (cathode composite without sulfur). The corresponding cyclic voltammogram (Fig. 2a) shows a cathodic peak at a potential bellow 0.8 V versus platinum reference electrode. Once polysulfide solution of Li_2S_8 and electrolyte are used, another cathodic peak appears in the potential range between 2.25V and 1.5V (Figure 2a). The height of this cathodic peak is in correlation with the concentration of Li_2S_8 in the electrolyte. As we showed in our previous work, the correlation is linear and in some standardized conditions it can be used for comparison of different systems. However, it cannot be used for the determination of overall quantity of polysulfides that has diffused away from the cathode composite. The cumulative (integrated) charge that corresponds to the area in the selected potential range corresponds to the reduction of only approximately 2% of polysulfides added to the electrolyte solution. Using the blank cell, we further observed that most of the polysulfides were reduced in the first scan, as shown in Figure 2b. The integrated charge in higher cycles corresponds to less than 5 % of the charge obtained from the first scan and it can be correlated to the polysulphides diffused from the rest of separator area. Typically we found a yellow to white precipitate in the separator which was in the vicinity of the working electrode (nickel or stainless steel wire), while the wire was not covered by any visible layer.

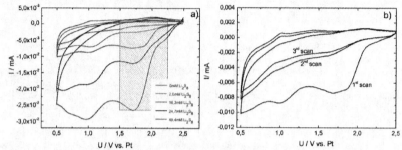

Figure 2: a) cyclic volmmograms for electrolytes in the blank battery configuration with different standards (different concentration of dissolved Li_2S_8) and b) consecutive three cyclic volmmogram scans for the same blank battery with standard electrolyte containing 24.7 mM Li_2S_8.

Application of the modified 4-electrode cell in combination with a real battery (cathode composite containing sulfur) confirmed our expectation that we can detect polysulfides during the battery operation. Cyclic volammograms were collected after every 2 hours of galvanostatical discharging in the first cycle (Figure 3).

The measured cyclic voltammograms showed evolution of the soluble polysulfides which can be scaled with the intensity of cathodic peak at approximately 1.8 V. Note that a remarkable change is only observed in the cathodic peak at 1.8 V while the second cathodic peak (at 1.3 V) remained relatively constant; the origin of the latter is not clear at the present state of the work. The maximum solubility has been detected after a change in the composition for $\Delta x=0.2$. During the next two steps it remained relatively high. Here we need to consider that we might need to reduce shorter polysulfides (Li_2S_6 and/or Li_2S_4) chains which can consume less charge that Li_2S_8 available in the beginning of the discharge curve. Later during the discharge, the amount of charge used for the reduction of soluble polysulfides becomes smaller which may be ascribed to a smaller diffusivity of shorter chain polysulfides and a smaller quantity of charge required for the reduction of short chain polysulfides (i.e for the reduction of Li_2S_3 and Li_2S_2).

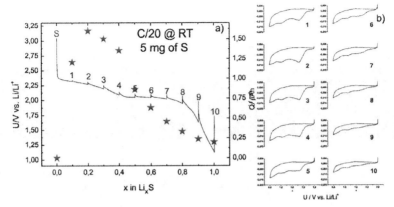

Figure 3: a) Electrochemical behavior during first reduction of Printex XE2/S composite and b) corresponding cyclic voltammograms measured after the change in composition of Δx=0,1. Red stars in (a) correspond to the charge obtained by integration between 2.25 and 1.5V in measured CV's.

As proposed in our previous work, various chemical environments with different electrochemical behavior showed differences in the potential region between 2.25 V and 1.5 V (changes in the integrated charge). Observed unique responses can be explained by knowing the physical and chemical properties of components used in the electrochemical cell. For instance in Figure 4 we show the starting CVs from four different chemical environments and the CVs at maximum solubility (the highest intensity of the cathodic peak at 1.8V). As expected, the battery in which we used the electrolyte containing dissolved chemically synthesized Li_2S_8 showed the highest response (Figure 4a – red curve) in the first scan - even before we started with the battery (the concentration of dissolved polysulfides was about 25 mM). Other examples presented in Figure 4a show only the cathodic peak at 1.3 V and we can observe the pronounced difference between the two electrolytes (for the EMS electrolyte, the cathodic peak was observed at potentials that were for several tens of milivolts higher than in the case of the TMS electrolyte; however, one also needs to stress that the shape of CV in this region is different for mesoporous carbon compared to carbon black). Figure 4b shows cyclic voltammograms obtained during the sequenced scans, as shown in Figure 3. In this case, as presented in Figure we observed the cathodic peak at approximately 1,8V however not in all examples at the same composition (Δx). Note that the highest integrated charge, as well the highest cathodic peak, was obtained with the battery where we added chemically synthesized polysulfides into the electrolyte. Surprisingly, a very high integrated charge was also obtained in the case when we used EMS as a solvent for the electrolyte. Here we need to add that this CV scan was measured in the middle of the cycling curve (not shown in this work) while using the TMS solvent for the electrolyte solution we observed the maximum intensity peak during the first plateau at 2.4V. That phenomenon can be probably explained with different viscosities of electrolytes. The difference in the results obtained with mesoporous carbon if compared to the ones obtained with a carbon black substrate (Printex XE2) is more or less expected; however, it is important that also in the case of mesoporous carbon we detected polysulphides that diffused away from the cathode composite.

Examples presented in this work show clearly that the use of modified 4-electrode Swagelok cell can help to better understand the electrochemical behavior (capacity fading and efficiency) of Li-S battery configuration. For example, recently we have shown that when oxide based substrates were used as a substrates for sulfur impregnation', the capacity fading was slower than in the case where mesoporous carbon was used. However, the formation was comparable in both cases. This unexpected result was explained by the use of 4-electrode Swagelok cell where we detected a much lower cathodic peak during the operation of battery containing oxide based substrate in the cathode composite. The reason for the smaller detected amount of polysulfides was not due to the more efficiently confined sulfur in the pores of SBA-15 but due to the weak bonding between pores and formed polysulfides. Modified 4-electrode Swagelok cell can be used as a reliable analytical cell for Li-S batteries where we can quantitatively determine the amount of polysulfides that diffuse away from the cathode composite.

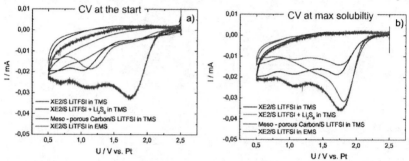

Figure 4: a) CV scans of different Li-S batteries obtained before galvanostatic dischargeand b) CV scans obtained during the sequenced scans measured during galvanostatic discharge where we observed the highest peak at 1,8V versus Pt electrode.

The second in-situ analytical technique presented in this work uses the property of polysulfides having different colors. In this way, UV-Visible spectroscopy becomes an effective way of detecting polysulfides, both quantitatively and qualitatively. Namely, polysulfides with different chain lengths reflect back (after absorption) different amount of energy when they are exposed to UV-Visible light. Thus, shorter chain lengths reflect back light with shorter wavelength and polysulfides with longer chain length reflects back light with a few tenths on nanometer longer wavelength. The color and consequently the position of spectra is a function, besides the chain length, also of the alkali metal in polysulfides and the medium where polysulfides are dissolved, while the intensity of the color is a function of the concentration. Lithium polysulfides dissolved in 1 M LiTFSI TMS electrolyte showed different colors going from reddish-brown for the dissolved long chain polysulfides to green-yellow color for short chain polysulfides.

UV-Vis spectra presented in Figure 5 were obtained in-situ during the galvanostatical discharge and charge of a Li-S battery that was mounted in UV-Vis spectrometer. Spectra were measured in the sequence of every 15 minutes (sequence of spectra changes from green to violet color in Figure 5). Clear reflection from the black surface can be observed in the beginning with no reflection in the visible light region. During discharge, polysufides were formed and the shape of UV-Vis spectra changed due to non-transparent nature of separator once it contained dissolved polysulfides. These changes are continuous until the maximum concentration and further discharging leads to the formation of the shorter chain polysulfides which differ also in the shape of UV-Vis spectra. Interestingly, during charge the changes are less pronounced since polysulfides remain in the separator and - following colors from green through pink to violet - one can see the polysulfide shuttle mechanism. In the start of discharging the position of the derivative peak in the visible light region is below 500 nm which suggests the presence of short chain polysulfides. With charging it shifts to higher wavelengths which could be interpreted as oxidation of short chain polysulfide to long chain polysulfides, but at the end of the charge we detected the formation of low chain polysulfides (presence of the violet peak in the derivative curve at wavelengths bellow 500 nm in Figure 5d). The most reasonable explanation for this phenomenon seems to be the polysulfide shuttle. Namely, long chain polysulfides formed during charging can diffuse to the lithium anode where they are reduced back to short chain polysulfides. Furthermore, polysulfide shuttle mechanism is additionally supported by the fact that the cathode composite used in this study was made of carbon black and sulfur. In any case, these results suggest that UV-Vis can be used as an analytical technique that can give some insight in the Li-S battery cycling behavior in the early stages of the battery cycling.

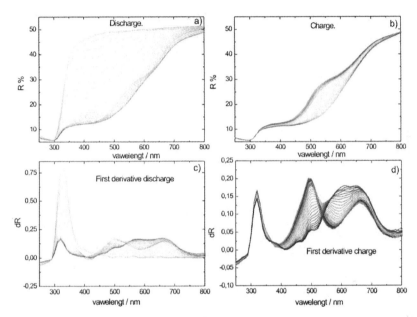

Figure 5: a) and b) UV-Vis spectra obtained during discharge and charge of Li-S battery in the first cycle (cathode composite was made of Printex-XE2 and sulfur). First derivatives of measured spectra are shown in c) and d).

From the investigation performed so far, it seems that the use of the first derivatives of UV-Vis spectra is particularly convenient. This way, one can compare quite reliably various spectra at the wavelength of between 300-800 nm. Figure 6a shows the spectra of standard solutions (10 mM/L) with different polysulfides Li_2S_n (where n=8, 7, ..., 2). TFSI dissolved in TMS did not show any absorption at wavelengths from 800 nm to 350 nm, while as expected different polysufides dissolved in the same electrolyte showed remarkable differences in color. From Figure 6b we can see that there are two sets of derivative peaks that are shifted with a change in the composition of polysulfides. The set of peaks observed in the wavelength region from 300 to 400 nm shows a high dependency on the concentration and it disappears at high concentration of polysulfides in the electrolyte solution. The second set of peaks - in the wavelength range from 400 to 600 nm - is less dependent on the concentration and consequently can be selected for a reference which can helps one to understand the processes in the battery. Here we show, in Figure 6c, selected UV-Vis spectra measured during discharge of Li-S battery and the corresponding derivative patterns which are shifted to enable a better view. First of all, if we compare derivative of starting UV-Vis spectrum with derivative of TMS/TFSI spectrum, we could not see any contribution from additional species from the battery. During the discharge different peaks can be observed in the derivatives of UV-Vis spectra. Surprisingly, after 5 consequent scans we can determine from the derivative the presence of short and long chain polysufides. However, it needs to be carefully checked how low concentration influences the shape of UV-Vis spectra and its derivative. With further charging (each 5 spectra correspond to the change in the composition of 0.12 mol of lithium) we can see the evolution of peaks that can be ascribed to longer chain polysulfides (like Li_2S_8 and Li_2S_6). In the latter stages of charging, the formation of shorter chain polysulfides (like Li_2S_3 and Li_2S_2) can be observed. Here we need to emphasize that all presented spectra correspond to the composition found in the separator since the geometry of the cell enables measurements close to the quartz window surface.

The proposed analytical techniques turned out to be capable of determining, either quantitatively or qualitatively, the polysulfides that diffused away from the cathode composite. Both features of this method are extremely helpful for the study of the dissolution mechanism, as well as for the detection of polysulfides in the separator. Last but not least, reliable and accurate detection of polysulfides in the separator has is important already in the early stages because it can point out to possible problems with polysulfide shuttle mechanism.

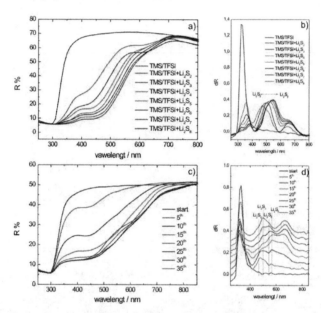

Figure 6: a) and b) UV-Vis spectra of standard solutions (10mM/L) with different polysulphides Li_2S_n (n=8, 7, 6, ..., 2) dissolved in 1 M LiTFSI/TMS electrolyte and their first derivates and c) and d) UV-Vis spectra and respective first derivates during the Li-S battery discharge (only every 5th spectra are shown).

CONCLUSIONS

In this work we presented two different in-situ analytical techniques which are capable to detect quantitative and qualitative differences between different polysulfide species in the electrolyte. Both proposed techniques can detect polysulfides diffused away from the cathode composite. Detection of polysulfides with modified Swagelok cell resulted quantitative charge of reduced polysulfides on working electrode of the cell which is built in between classical battery sandwich. Integration of the charge in the potential range from 2.25V to 1.5V versus platinum electrode is a function of the concentration of polysulfides in the electrolyte solution and this can be used for detection and determination of amount of polysulfides that diffused away from the cathode composite. Within standardized conditions, the proposed analytical technique can serve as a tool for the comparison of different chemical environments and cathode compositions/architectures used in Li-S battery. Use of UV-Vis spectroscopy returned more qualitative determination of polysulfides and it can suggests which species (polysulfides) have higher mobility in the different chemical environments. Combination of both analytical techniques with a classical galvanostatic cycling could lead to better understanding and with consequently faster tuning of Li-S battery properties.

ACKNOWLEDGEMENT

The authors acknowledge Slovenian Research Agency and ALISTORE European Research Institute (ERI) for the financial support.

REFERENCES:

[1]P.G. Bruce, S.A. Freunberger, L.J. Hardwick, J.-M. Tarascon, *Nat. Mat.*, **11**, 19-29 (2012)

[2]R.D. Rauh, K.M. Abraham, G.F. Pearson, J.K. Surprenant, A.B. Brummer, *J. Electrochem. Soc.*, **126** 523(1979).

[3]H. Yamin, A. Gorenshtein, J. Penciner, Y. Sternberg, E. Peled, *J. Electrochem. Soc.,* **135** (1988) 1045.

[4]V.S. Kolosnitsyn, E.V. Karaseva, *Russian J. of Electrochemistry*, **44** (2008) 548.

[5]X. Ji, K.T. Lee, L.F. Nazar, *Nature Materials*, **8** (2009) 500.

[6]R. Dominko, R. Demir-Cakan, M. Morcrette, .-M. Tarascon, *Electrochem. Commun.*, **13**, 117-120 (2011).

[8]R. Demir-Cakan, M. Morcrette, F. Nouar, C. Davosine, T. Devic, D. Gonbeau, R. Dominko, C. Serre, G. Ferey, J.-M. Tarascon, *J. Am. Chem. Soc.*, **133**, 16154-16160 (2011).

[9]J.R. Akridge, Y.V. Mikhaylik, N. White, *Solid-State Ionics*, **175** (2004) 243.

[10]W. Zheng, Y.W. Liu, X.G. Hu, C.F. Zhang, *Electrochimica Acta* **51** (2006) 1330.

[11]L.X. Yuan, J.K. Feng, X.P. Ai, Y.L. Cao, S.L. Chen, H.X. Yang, *Electrochem. Commun.* **8** (2006) 610.

THREE NEW APPROACHES USING SILICON IN THREE VALUABLE ENERGY
APPLICATIONS

John Carberry
Mossey Creek Energy
Jefferson City, TN 37760
johnc@mosseycreekenterprises.com

ABSTRACT
    The new availability of economical industrial volumes of nano and micron sized silicon
processed free of contamination and oxidation even at the native oxide layer enable new methods for
using silicon in three valuable energy applications; Thermo Electric Generators and Coolers based on
Porous Silicon, Net Shape Photovoltaic Wafers, High Thermal Conductivity Epoxy Molding
Compounds, as well as others.

THERMO ELECTRIC GENERATORS AND COOLERS BASED ON POROUS SILICON:
    But for silicon's high thermal conductivity (K), it is otherwise the ideal thermoelectric material:
it's very low coefficient of expansion (CTE), its high Seebeck Coefficient, its ability to be tailored for
specific semiconductor and electrical and thermoelectric properties, its capability to operate at high
temperatures, and its low relative cost make it the most attractive tool for harvesting waste heat as
electricity and cooling in refrigeration and air conditioning if its K value could be controlled. So the
key issues for success for silicon in this most valuable application is its high thermal conductivity of
149 W/MK.
    In an exciting recent development in quantum physics, announced by multiple investigators,
phenomena have described a quantum size effect achieved through nano structuring which provides a
means to tailor silicon structures for exceptionally low thermal conductivity, as low as .4 W/MK, a 350
times reduction in this denominator in the thermo electric power factor equation. The phonon in silicon
has been characterized as being on the order of hundreds of nanometers and being blocked by nanostructures on the order of tens of nanometers. This breakthrough brings to us a path for using silicon to
make thermo electric generators and coolers that can change the world.
    Many of the recent efforts and developments in this regard, many funded by DARPA and the
DOE and NASA in the US, have focused on nanowires and MEMS, which have brought forward
announcements confirming exceptionally high power factors with very high efficiencies in converting
waste heat to electricity. Unfortunately, these structures are expensive and cannot be made in the thick
cross sections required to maintain a large unaided or largely unaided Δ T. Many of the results
reported used aggressive heat exchange apparatus to maintain a high Δ T. In this paper we introduce an
approach which is industrializable and highly economical.
    The thermoelectric figure-of-merit, ZT, for a thermoelectric material (TEMat) is a
dimensionless measure of its efficiency. Z is calculated by multiplying electrical conductivity (σ) and
Seebeck Coefficient (S) squared and dividing by thermal conductivity (K), or:

$$Z = S^2 \sigma / K \tag{1}$$

    ZT is calculated by multiplying Z with absolute temperature (in Kelvin). To achieve a high
power factor, it is therefore desirable to have a TEMat with low thermal conductivity, high electrical
conductivity, high Seebeck coefficient, and with a very low coefficient of thermal expansion (CTE)
enabling a high temperature operating capability (i.e., a sustainable temperature difference across its
structure or ΔT capability).

11

Exploiting a TEMat's ZT is more than materials science. Successful exploitation will need to combine brittle material engineering practices as TEMats, as a material class, are very brittle (i.e., low fracture toughness). A prerequisite to exploiting a TEMat high temperature capability and its ZT is it must be able to also mechanically withstand a large mostly unaided ΔT in service. This in turn results in a need for the TEMat to have a minimum coefficient of thermal expansion (CTE) and maximum tensile strength (S_{Ten}). Lastly, from a perspective of size, a larger TEMat component or "leg" will promote the ability to achieve a larger ΔT (presuming it does not mechanically fail); this is an important issue for achieving cold temperatures too.

Incumbent technologies offer little hope of making low cost thick structures able to operate at high temperatures with high-unaided ΔT and attractive power factors. Traditional and new approaches to making thermoelectric generators (TEGs) are all flawed by fundamental and seemingly intractable challenges, such as high cost, high CTE, limited to thin planer structures, low S, low electrical conductivity, low mechanical strength, or use of rare and costly materials, or combinations thereof. Many of those same issues limit the ability to achieve colder temperatures with thermoelectric coolers (TECs).

For instance in their 2011 paper [1], Xingzhe Wang et al teach that silicon's thermal conductivity is insensitive to grain size till the grain sizes are reduced to quite a bit less than a micron, and then falls precipitously from about 600 nm to 5 nm with thermal conductivity falling to less than .4 W/Mk. But they only address a "bulk" material. They do not describe methods or sources for a silicon bulk material with a grain size in the range of a few or tens of nanometers, but conclude that only by reducing the grain size can one obtain silicon with very low thermal conductivity.

An ideal pathway for making thermoelectric devices would include a way to obtain porous non bulk nano sized equiaxed silicon grains that could be formed into robust large shapes with large cross sections and a nano structured morphology, so to achieve or promote a low CTE, a low value for K, very high values for electrical conductivity, gain high S values, and high operating temperature capability.

Silicon's very strong exothermic reactivity with oxygen has made it heretofore very dangerous to mill into micron and nano sizes, and in fact many lives have been lost and many injuries incurred in silicon milling explosions. In many cases this has reduced or eliminated the economical use of silicon in many applications. This is because silicon is one of the most reactive elements in the periodic chart, proven by the fact that in nature one finds the oxides and silicates, but virtually never silicon! One will find the industrial availability of finely divided un-oxidized silicon to be very rare and expensive. For instance at this time in June 2012, the only source I can find for finely divided silicon with a surface area of 80 meters/cc and a D50 of 50 nanometers is sold for 19 EUR a gram at 99% purity. (PlasmaChem 2011/2012 *Catalogue for Nanomaterials and related products*)

Our innovation is based on the clever combination of five disparate but proven concepts, which in combination bring about a breakthrough in thermoelectric efficiency. These are the:

1) Proven ability to economically mill silicon to nano sized equiaxed grains without contamination or danger of explosion;
2) Insight into the decoupling of the electron and phonon through the use of phonon interference at very small interfaces created by this process;
3) Protection of the silicon powders from oxidation between milling and the sintering step, noting that the native oxide layer acquired after sintering will not affect electrical resistivity of the body, but that oxidation of the surfaces of the grains before sintering is likely to increase electrical resistivity.
4) Use of boron and compensated phosphorous doping to create very high electrical conductivity and effective P and N type doping without increasing thermal conductivity; and,
5) Ability to form the nano-sized silicon powder into porous, strong, large, complex shapes as required in both P and N type TEG pellets with ZT values as high as 6.

In this perspective we address the recent developments reported in the literature with regard to engineering low k. One must show how the addition of these new insights to traditional models for K in porous structures can be differently understood.   This is the enabler for MCE to optimize the engineering of nano structured morphology in silicon structures with very low K.

In looking at the models and literature for the thermal conductivity of porous structures and TEG, especially those working with nano wires and holey silicon, a few things become clear. Here is referenced some work which shows the combination of these insights which leads us to innovate in new ways from the current materials and silicon approaches, from nano wires and MEMS towards our new industrial method for making low cost high output TEGs. Here is highlighted the diverse elements and approaches that underlie this approach.

Traditional models for the thermal conductivity of porous structures show very good work done for many years, for example John B. Butt's January 1965 article [2] taught:

a) The models do not capture the empirical measurements very well, he writes in his summary "Although it would be a serious error to assume that the representation of what is happening is correct in each detail, presumably the concepts which are utilized by the method result in an appropriate overall evaluation of the process";

b) The difficulty is the use of these four terms and finding an accurate characterization of the structures being modeled and those being measured:

1. Gas phase conduction through macropores
2. Gas phase conduction through micro-pores
3. Conduction through micro and macropores in series
4. Conduction through the solid phase

These models do not account for the low k of nano structured porous silicon or these nano-wires or holey silicon or even the structures taught herein, and Butt's work did not address such nano phase materials and phonon interference at interfaces which has only been recently reported and described in the literature. Here one finds recent work focusing on phonon interference help explain the gap between the traditional models exemplified by Butt and the exceptionally low k being made and measured in current work.

Fang and Pilon [3]: "It has been established that the reduction of thermal conductivity in nano-composite materials is mainly due to phonon scattering by interfaces. The phonon-interface scattering is known to increase with increasing interfacial area concentration Ai, (M-1) defined as the surface area of interface per unit volume of nano-composite material". Fang and Pilon give three references, all recently reported as well, which reinforce this thesis: 1) A. Minnich and G. Chin [4] 2) M.S. Jeng, R.G. Yang, D. Song and G. Chen [5] 3)K.M. Katika and L. Pilon [6]

Lee, Galli and Grossman [7] reported: "The authors ascribed the sharp increase in ZT to the reduction in k and the phonon drag effect, the latter of which has been known to be small or negligible in previously studied TE materials."

Sabah et al of JPL [8] reported the making of small samples with a shaker mill using inert cover gasses, and dealing with intractable industrial problems with agglomeration and very broad particle size distribution, from 10 nm, to several microns with problems of agglomeration. Employing our '491 patent for milling in alcohol MCE do not have this agglomeration and can mill quickly and economically to very narrow particle size distribution thus achieving both the required doping to manage electrical conductivity and the very low thermal conductivity managed with phonon interference. Even with the shortcomings reported Sabah reported a ZT of nearly 7.

So why did Sabah et all fail to achieve a higher ZT? They did use the laboratory shaker mill to successfully mill to very small nanosizes, but then they hot pressed the silicon into a bulk structure and were for this reason only able to achieve a thermal conductivity of 7-8 W/MK. MCE is making a porous structure with thermal conductivity fully exploiting the quantum size effect of these nanostructures and therefore with an approach otherwise similar able to achieve a ZT of 6 or more.

Furthermore if one reviews shaker mill equipment available and also looks at the energy and infrastructure required to set up a large shaker mill, one concludes that in addition to the intractable problems of agglomeration, explosion danger and broad particle size distribution it is just not possible to industrialize a shaker mill; there is simply too much mass moving with too much momentum and shifts in momentum at too high a frequency to engineer the energy or the frame work. The attrition mill approach limits the energy to the shaft and the paddles and the media and the work and of course because everything is in liquid there is free movement of the work and media and no agglomeration.

Ci et al. [9] report several key observations:

- "The plot shows that the thermolectromotive force (ΔV) increase quickly when the temperature difference (ΔT) compared to room temperature increases" thus underlying the importance of high operating temperature";
- "The enhancement in the Seebeck coefficient is due to the change in the energy band while the pore wall becomes thinner. Therefore, Figure 7 indicates that a smaller thickness of the pore wall will increase the Seebeck coefficient notably";
- "However, doping can reduce the resistivity, and more work is underway in our laboratory";
- "The Seebeck coefficient is proportional to the thickness of the pore wall. This phenomenon is believed to be related to the change in the energy band, which affects the effective mass of the carriers. In addition, the thermal conductivity of silicon microchannel plates which has similar porosity to porous silicon is low."

Of course, as Ci et al. oxidized larger and larger cross sections of the wall thickness and removed the silica by HF etching, they were creating smaller and smaller cross sections, meaning they were creating stronger and stronger phonon interference at the interfaces. Since ΔT is a strong factor in tailoring a material for a high S, the low k resulting from this stronger and stronger phonon interference at the interfaces by this etching resulted in the very high S value they reported, from 466 up to as high as 1019 $\mu V/K$.

Therefore the model of phonon interference provides the missing additional terms needed to use Butt's model so to more completely model the very low k being measured, lower than anticipated by Butt's models without these additional terms.

So next we briefly address the idea that one can now dope the silicon with materials such as boron that can dramatically increase the electrical conductivity, in essence separating the terms describing k, being that of the electron and the phonon. In this case the phonon is not able to efficiently pass through the barrier at the interface, while the electron passes without any resistance or interference. This gives us the opportunity to do doping of our silicon to increase electrical conductivity without increasing the thermal conductivity, essentially decoupling of the duality of the electron and phonon terms.

From recent literature one finds that this opportunity has been recognized. Pengliang et al. [10] report the ability to dope with boron to increase electrical conductivity to a reported optimal 133 $W^{1}cm^{-1}$ while controlling k and S so to achieve a ZT of 4.7.

The issue has been addressed in the area of nano wires, porous silicon and holey silicon. Researchers have proven that one can dope with boron and achieve the desired electrical properties, and reported measurements for ZT in the range of 1.2 - 4.7. A ZT greater than 1 enables a wide range of TEG applications, a ZT greater than 1.5 enables automotive refrigeration and the like, and a ZT of approximately 3 enables very rewarding economic industrial waste heat recovery.

An economical TEG "Leg" able to achieve these levels of efficiencies and economics offer a very attractive capital cost for recovering waste heat as electricity and functioning as an engine for removing heat for cooling.

The process we disclose also invents an object of great value when incorporated into application specific structures unique to these objects, the large legs, for a variety of applications and examples such as:

1) Solid state cooling applications include:
   a. Residential refrigeration and air conditioning (20% of residential electricity charges represent the fridge) where electricity use can be reduced by 50% or more,
   b. Retail and commercial and industrial cooling and freezing and air conditioning applications,
   c. Mobile cooling for air conditioning in cars, trains, buses and the like, transport cooling for trucks, containers and the like where not only is power consumption reduced but the weight of the unit can be reduced by as much as 75% which commercial carriers would of course replace with additional revenue bearing freight.
   d. Cooling of power devices such as semiconductors, electronics and the like.
   e. Industrial cooling for process controls
   f. Chilling of water for marine aquariums

2) Waste heat recovery by conversion to electricity applications include:
   a. Efficient recovery of the more than 500 centigrade heat flowing at the exhaust manifold brings the capability to produce thousands of watts of electricity that can dramatically improve efficiency and performance of cars, trucks, trains, boats and planes:
      i. It can produce enough electricity to replace the alternator, thus reducing the drag on the engine and removing an expensive unreliable component system;
      ii. It can produce enough electricity to power lower weight more reliable electric motors to move oil and radiator fluids further reducing drag on the engine;
      iii. In combination with thermo electric air conditioning it can reduce the drag on the engine further and reduce the energy for air conditioning further;
      iv. It can produce enough electricity to also charge film capacitors, which would be able to carry a large burden in propelling the vehicle during acceleration profiles from a start or in increasing velocity in acceleration, further reducing horsepower required;
      v. In all these ways it is possible to calculate increases in fuel mileage as much as 100% or more by reduction in horsepower and increases in efficiencies, while maintaining drivability and quality of experience.
   b. Industrial waste heat in the US is equal to 68 Quads of energy (68 X10^{15th} BTUs). Most waste heat produced in large volumes over 600 centigrade can be economically converted to steam and electricity. But heat in the range of up to 600 centigrade will be most efficiently converted and recovered as electricity through our invention. With such high Δ T, a market not addressed by all other technologies due to several factors including high CTE, the large legs become so very valuable, so we describe an invention where the long legs herein described are incorporated into modules able to convert very high temperatures on one side into electricity while maintaining a high Δ t.
   c. Co-processing of power in many applications is enabled. For instance, solar cells are designed to absorb as close to 100% of the sun's energy as possible and their backs reach temperatures as high as 160 centigrade. The hot side of all the cooling devices described above are sources of heat than can be so converted. Aerospace and

defense applications often need to cool devices but also convert heat to power where power is so expensive, for instance in airborne and space applications.

3)   The materials in the family of compounds using tellurium, and other such materials, suffer from ills such as the fact that China controls the bulk of the world's supply and has started to control it for strategic sovereign business interests; supply is now short and prices have skyrocketed; it has very high CTE values, limiting operating temperature and size; its thickness is limited because of cost, CTE and manufacturing processes; it also has a relatively low S value;

4)   Silicon nano-wire and micro-channel structures made from MEMS and semiconductor processes suffer from ills in that they can only be made in very thin cross sections; they are hard to dope; they are very expensive and they have limited operating temperature range.

5)   Recent work with micro-channels and nano wires, and well established characterization of "porous silicon", have highlighted the fact that when the structures of the silicon include pores, spaces, contacts measured in the tens of nanometers, phonon interference at these interfaces trump traditional models for thermal conductance providing low k as low as 0.1 to 1 W/mK.

6)   Additionally, recent reports that doping of silicon with boron provides a means to engineer very high electrical conductivity while using these nano-structured morphologies to limit k while maintaining high S.

7)   While the micro-channels and nano-wires have shown one can achieve these characteristics of low k and high electrical conductivity while maintaining high S, they still suffer from very high costs, difficulty to make complex net shapes, and extreme difficulty of making large cross sections necessary for unaided high ΔT.

Silicon has the advantage of a high Seebeck Coefficient, a very high capability for ΔT, the ability to be tailored as a semiconductor, and now the ability to achieve very low thermal conductivity.

The most common thermoelectric materials used to date are typified by bismuth telluride. (BiTe). These materials suffer from several limitations when compared to silicon, now that silicon can offer a low thermal conductivity:

Property	Unit	BiTe	Silicon
CTE	PPM	18	3.8
Thermal Conductivity	Watts/MK	1.5	0.5
Seebeck Coefficient	μV/K	50-80	300-1100
Operating Temperature	Degrees C	100-150	600-1100

Mossey Creek Energy's US Patent 6,638,491 B2 [11] provides a unique, safe, low cost, industrialized process for milling silicon to a few hundred nanometers. This has provided the platform from which the author has investigated and from which we are developing a game changing thermo electric technology suite. We believe these offer highly valuable solutions in these energy applications relating to the Seebeck physics. For instance:

Milling Percentiles safely using the '491 patent free of oxygen:

10% = 0.103   20% = 0.116   30% = 0.134   40% = 0.170   50% = 0.318

We believe, given the lack of availability of pure, unoxidized finely divided silicon, most all investigators have taken a pass on this approach, unless as in my case, one is skilled in the processing of same. This perhaps is a first proof of bringing insights one of ordinary skill in the art would not be able to bring to bear. For instance if the wave form of the phonon in silicon at 400 degrees K is 220 nanometers, a ten nanometer grain boundary interface will more or less eliminate thermal conductivity at that interface. The phonon will be internally reflected and thereby, more probably, find an opportunity to become an electron. This is our design.

Our invention introduces an innovative industrial method for processing suitable silicon precursors into nano sized grains, essentially equiaxed, which can be formed into large net shape simple or complex thick shapes and sintered into structures containing these nano structured morphologies while containing dopants, for instance boron, that provide high electrical conductivity and S in such a way as to preserve high carrier concentrations and long carrier lifetime. A very important element for success is protecting the silicon surfaces before and during sintering from any oxidation as a small amount of oxidation in these very small grain boundaries can cause very large increases in electrical resistivity.

The result is a low cost, very efficient (high-valued ZT), high output thermoelectric engine for harvesting waste heat as voltage and providing powerful engines for heating or cooling. In this approach we address a method using my '491, patent described below, and we believe this is currently the optimal method for obtaining precursor small nearly equiaxed small nano sized silicon grains.

We note that when making a bulk material, which is by definition fully or nearly fully dense, the grains conform to each other, and therefore, referencing Wang's work in my approach, what I call the "grain boundary interface area" is equal to the surface areas of the grains themselves. Therefore to conform to Wang's excellent and helpful curve to achieve thermal conductivity equal to or less than competing materials such as are available in the current thermoelectric world one would have to control grains in the bulk materials to sizes in the range of 5-20 nanometers, a most formidable challenge as all in the field would attest, as would the lack of such a published commercial solution.

We observe that our invention, in which we make a porous structure, which we are able to "dry press" under high pressure into strong structures without using binders, and this allows us to "sinter" the grains without densifying them. This results in a "grain boundary interface area" that is far less than the surface areas of the grains. Instead the grain boundary interface area is defined by the "Hertzian" contact area, which is only about 1/40th of the diameter of the grains. This means that a D 50 milling of silicon resulting in processing of about 300 nanometers for example would give grain boundary contact areas of about 8 nanometers, which Wang's formula predicts would have a thermal conductivity of less than .4 W/Mk. In this approach, we reduce the necessity for making the smallest grains and achieve excellent control and costs by using "larger grains" to achieve the smallest grain boundary interface area in large bulk structures we need for high unaided ΔT, thus optimizing phonon interference and reduced thermal conductivity. We incorporate this approach to making these silicon grains for making this porous structure using the '491 patent, we also incorporate any and all other methods of obtaining such precursors.

We have employed the '491 patent to prepare silicon, sintered TEMat billets have been produced with K values of .5-3 W/mK at 500°C and S values as high as 470 µV/K. From these perspectives my innovative work has brought these insights and developments:

1) Small nano structure morphology in silicon, where the interfaces are in the range of a few to tens of nm between small structures, of wires, micropores, MEMs structures, or my milled silicon grains, will have phonon interference at the grain boundaries and not interfere with electrical conduction. This is the key driver in lowering silicon's otherwise high K without concomitantly reducing its electrical conductivity;

2) We have analyzed this and predict the silicon grains are bonded together with a combination of amorphous silicon, silicon oxide and hydroxyls and silicon itself, which is also, in addition to the nano sized grain interfaces, a powerful phonon transfer blocker;

3) This particle size portrait of silicon milled according to the '491 patent to a D50 of 300 nm, shows industrial ability to make these precursor powders as small and as narrowly

distributed as required. Through this patent this requirement for preparing the precursor powders by milling the silicon is successfully addressed. Additionally, subsequent patents develop the ability, in the milling process, to add the boron and or phosphorous for doping for electrical conductivity and making N and P type semiconductors. The boron and phosphorous will completely diffuse at 1370°C, providing lower cost of the raw materials.

4) Doping with boron promotes high electrical conductivity. Our work and the work reported by other investigators shows that one can dope with boron to very low levels of electrical resistivity and that the Seebeck Coefficient will drop to about 260 μV/K, at which point it becomes insensitive to further increases in electrical conductivity.

5) This can be done for micro-channels, MEMS or nano wires, but these cannot be large, thick, net shape for complex shapes, and low low cost.

6) We can also thus prepare silicon alloyed with other materials, for instance germanium, typically in the range of 20% germanium and 80% silicon, which will aid milling by increasing brittleness, and make a porous structure of alloyed and doped silicon alloyed with germanium thus increasing temperature operating range and increasing the power factor.

7) Most all thermoelectric materials extent today are not able to make thick "Legs" meaning that even with a very low thermal conductivity, it is challenging to maintain a large Δ T since without space it is difficult to avoid thermodynamic equilibrium. A large leg would support the thesis that one could use a material with high thermal conductivity and still achieve a large Δ t, and that with a large leg and a low thermal conductivity one would achieve the highest Δ t. Current TMAT legs are measured in the range of hundreds of microns, so we are describing and inventing a technology where our legs are an order of magnitude larger, so in the range of many mm or tens of mm, our current approach is a leg 25 mm tall. This is possible with the silicon we are making because it has such a small CTE compared to other materials and is strong, and economical to make in these larger leg structures. This provides much superior performance and is a key feature of our invention, the ability to make much larger legs with low thermal conductivity sustaining a very significant improvement in ΔT.

8) Therefore the approach taught herein is a novel, unique, and unexpected way to make economic TEG structures which:

   a) Using the '491 patent process I can make very small grain to grain interfaces. For example, the silicon powder below can be industrially milled and pressed into pellets and this represents an optimized size distribution at about 300 nm.

   b) These small grains are then pressed or cast or formed into the desired shapes using traditional and established ceramic fabrication technologies;

   c) When sintered in Argon, these nm-sized powders can be formed into strong structures which also have a low CTE ($\sim 4 \times 10^{-6}/°C$);

   d) These powders can be doped to control the electrical conductivity and to construct P and N type semiconductors in two ways:

      i. "Compensated" solar grade silicon or electronic grade silicon can be used with a base doping of boron to create desired K, and in the case of the N type an appropriate amount of phosphorous, for instance, can be added to compensate for the boron and provide for the carrier behavior desired;

      ii. Since we are milling to such small sizes we can also add the dopants during milling and rely on the diffusion at sintering temperatures, which will happen efficiently, thereby using materials such as pure Siemens grade or upgraded solar silicon as a starting material.

    iii. By sintering boron-doped silicon to make P type semiconductors, which has been milled to above referenced particle size and distribution per Carberry's '491 patent, that these results have been achieved, and in future work compensated pellets doped with phosphorous over boron will duplicate these results in N type semiconductors.

We therefore describe a process and an object wherein a very high ZT can be created in a TEG or TEC using silicon that will have the capability to be made inexpensively in many shapes large and small and have properties with these estimated ranges:

Seebeck coefficient (S):	300 - 1100 μV/K,
Electrical conductivity (s):	100 - 1000 $W^{-1} \cdot cm^{-1}$,
Thermal conductivity (k):	0.1 - 7 W/mK -100°C to +600°C,
Operating temperature:	-100°C to +1100°C,
CTE:	3 - 5 x 10^{-6}/°C.

NET SHAPE PHOTOVOLTAIC WAFERS

Review of Net Shape Solar Silicon Wafer Developments by MCE and ORNL:

Current silicon wafers for PV are made in crucibles which are costly, cause massive contamination and require expensive wire sawing that micro-cracks the wafers. Currently the Chinese 350% capacities overbuild in world production and heavy subsidization are threatening the very survival of all European and US PV solar cell makers. The only obvious pathway going forward is to reduce the costs of making the wafers and increasing their output. We believe that this can only be accomplished by making a high quality, high purity, unstressed net shape silicon wafer. The parameters for approaching the solutions for this problem are several, and incumbent in the current processes:

1) The absence of oxygen is a very key parameter;
2) It is perhaps impossible to create a processing environment so free of contaminants that speed of processing is not very attractive;
3) The fused silica crucibles used in polycrystalline and ingot making have about 350 ppm dissolved iron which is zoned refined as the fused silica precipitates as cristobalite and then flows freely as the viscosity drops as the iron content rises and that iron and most metals and transition metals especially diffused physically with great rapidity through a molten mass of silicon. Therefore, we understand that as we look at the problem of grain boundaries and recombinations we have become interested in the question as to are the grain boundaries so much a problem as what is in them? For instance it is very clear that the iron and metals in solution in the molten silicon get zone refined in the grain boundaries and that recombinations are found to concentrate in these areas of grain boundary contamination;
4) Wafers made by both single and polycrystalline find oxygen is at 25 ppm or higher in both;
5) Particularly in single crystal the oxygen tends also to flow in "striations" along the draw, creating great boundaries for recombinations;
6) It is important to understand the capillary forces at work in our situation because one will need to work with something that will not react with the silicon but which will wet it, or will not wet it but use such strong capillary forces in the geometry of our tools such that we can quickly fill the capillary spaces in conflict with the non or low wetting behavior to make a very thin, 160 micron or thinner, wafer quickly:

Capillary forces can powerfully assist us in this regard. Having said this the wetting angle of SiC, Dense Graphite 1.8grams/cc, and Si3N4 are respectively an attractive 0, 20-25 and 15-23 degrees. If we are not moving much and we are fast in and out with our melt, we should be able to use a capillary space to capture and make a thin net shape wafer. Having said this, one would note that we should be able to focus on melting the silicon purely and realizing these capillary forces through the meniscus height and capillary force calculation for silicon. So, if we consider that the viscosity of silicon at 1550 is about .7 which is about what water is at room temperature then this analysis must be applied to understand why we need to use tooling incorporating capillary spaces. The height h of a liquid column representing the power of the capillary force for a given system is given by:

$$h = \frac{2\gamma \cos\theta}{\rho g r},$$

(2)

Where:

- γ is the liquid-air surface tension (force/unit length),
- θ is the contact angle,
- ρ is the density of liquid (mass/volume),
- g is local gravitational field strength (force/unit mass), and
- r is radius of tube (length).

For a water-filled glass tube in air at standard laboratory conditions:

- $\gamma = 0.0728$ N/m at 20 °C,
- $\theta = 20°$ (0.35 rad),
- ρ is 1000 kg/m³, and
- $g = 9.8$ m/s².

For these values, the height of the water column is

$$h \approx \frac{1.4 \times 10^{-5}}{r} \text{ m.}$$

(3)

Therefore:

- For a 2 m (6.6 ft) diameter tube, the water would rise an unnoticeable 0.007 mm (0.00028 in).
- For a 2 cm (0.79 in) diameter tube, the water would rise 0.7 mm (0.028 in), and
- For a 0.2 mm (0.0079 in) diameter tube, the water would rise 70 mm (2.8 in)
- For a .02 mm diameter tube the water would rise 700 mm;
- For a .05 mm diameter tube the water would rise 280 mm;
- For a .1 mm diameter tube the water would rise 140 mm, if my calculations are correct.

.2 mm is of course 200 microns and a capillary force which would raise water 2.8 inches would obviously cause the silicon to enter into and stay in capillary confinement as a wafer on its side.

Having said this the capillary forces look completely adequate to capture and hold a silicon in the form of a wafer in the range of 400 microns down, if the silicon has melted pure and without surface and other contamination from SiO, SiC and such species. This underlies the nature of the investigations we are doing to confirm we can have pure silicon and can melt it without contamination.

7) We know that using the silicon nitride powder bound by colloidal silica Astropower obviously made wafers about 400 microns thick, a thickness necessary to overcome the surface tension and energy. We have recently observed that the coating needs to be preprocessed at about 1450

in nitrogen, which in the presence of so much carbon favors an exothermic reaction resulting in CO2 and Si3N4, which is much more favorable a wetting surface still being relatively non reactive. In effect the silicon nitride coating is relatively free of oxides.

8) It is clear that though the industry often starts with very pure silicon often 9 nines or purer that it is perhaps true that no one has ever made a wafer more than 4 nines pure. One wonders what the performance of a pure wafer might be. Such a wafer would need to be made quickly with very pure powder in a net shape in a very clean environment.

9) One can use inductive heating to heat graphite boats coated perhaps with the silicon nitride coating to heat to 1500 c in 10 seconds and cool quickly.

10) We have concluded that silicon is exceptionally reactive when molten and that processing speed is a very good thing since it reduces the time for reaction. We have also concluded that some materials are so reactive and are able to diffuse so rapidly into the molten silicon that they cannot be present in any form at all. These include slow diffusers such as oxygen and nitrogen because of their effect on viscosity especially; they form a skin that stops all melting flow. These also include iron and metals and especially transition metals which not only enter the melt very quickly but diffuse at astonishingly rapid rates and are poisons. They also include the semiconductor dopant materials, such as column three and five, boron and phosphorous and the like. Having said this, there appears to be nowhere on the periodic chart one can go and find non reactivity with molten silicon. Of course nature offers us a clues about this strict meta-stability of silicon in that one nearly never ever finds it in nature, it is an oxide, silicate and the like most always!

In this way we conclude that using the '491 patent to make a fine grained oxide free pure precursor silicon powder we can rapidly melt in specially prepared graphite tooling having a silicon nitride coating presenting capillary spaces for the thin pure net shape low cost wafer this application demands.

HIGH THERMAL CONDUCTIVITY EPOXY MOLDING COMPOUNDS

A number of applications for encapsulating semiconductor devices, inverters, film capacitors, motor elements and the like with epoxy molding compounds (EMC) are bringing a greater emphasis on the need for high thermal conductivity, which must by definition be contributed by the EMC filler. These include power semiconductors, semiconductors for ignition modules, LED lighting applications, inverters, capacitors and applications where device density also emphasizes the need for thermal management through higher thermal conductivity.

The typical materials used in designing and making encapsulating compounds for semiconductor devices are plastic molding compounds which typically have a low dielectric. These plastic molding compounds, which are composite materials, normally include epoxy resins, phenolic hardeners, mineral fillers, catalysts, pigments, and mold release agents.

Several important properties of the filler include the coefficient of thermal expansion (CTE), thermal conductivity, morphology, particle size, particle size distribution, effect of the filler on spiral flow and viscosity, electrical properties, and cost.

Traditionally, 40 years ago, typical mineral fillers included milled silica, quartz, and glass. During the past 25 years fused silica and fumed filler became dominant as they brought a low CTE and could be combined in particle size distribution that favored low viscosity flow at temperature, good spiral flow and high solids loading, as high as 80% or more. The key drawback of fused silica is the low thermal conductivity.

The value of achieving this can be found in many applications, many of which can have a critical impact on the CO2 foot print worldwide:

• LED lighting is limited mostly and uniquely by the requirements for thermal management. LEDs are semiconductor devices and their spectrum, lifetime and operating parameters are largely driven by operating temperatures at the junction which can only be controlled and optimized by the heat sink and the EMC. This is a very worthwhile endeavor as a long life broad spectrum LED could reduce energy for lighting worldwide from the current 1.2 TWh a year by as much as 95%

- BeO is still used in some power semiconductor applications, notably, the power ignition module. There is a strong desire to eliminate BeO because of its toxicity, cost and high CTE;
- Making all semiconductors more efficiently with respect to thermal management will extend life time, reduce energy required, and increase performance, all in ways that can dramatically reduce cost energy and the world's CO2 foot print.
- Inverters and film capacitors used in many applications, notably hybrid automobiles for instance, require very high thermal conductivity to optimize performance and weight to power ratios.

While the future may provide a low cost Aluminum Nitride powder, which can be seen to be motivated by the low CTE and high thermal conductivity and low dielectric constant, the high cost is today too high to make it an attractive candidate.

Filler Candidates

W/M*k	Thermal Conductivity	Specific Gravity	Est Cost USD/kg
Silicon	149	2.329	10
Silica	1.38	2.634	2
Alumina	18	3.99	5
Beryllia	330	3.02	800
Aluminum Nitride	280	3.26	400
SiC	4	3.2	40

Substrate Candidates which might be milled into powder for fillers:

	Specific Gravity grams/cc	CTE PPM	Thermal Conductivity w/M*K Theoretical	Thermal Conductivity w/M*K As made	Specific Heat J/KG*K	Resistivity Ohm/cm
BeO	3.02	8	330	265	1925	10 15
AlN	3.26	4.5	320	140-180	740	10 14
Al2O3	3.99	8.1	30	18	880	10 14

So here we can see that:

- Al2O3 while being inexpensive and a low dielectric, has a high CTE and a low thermal conductivity;
- BeO has a high thermal conductivity, low dielectric, but is toxic and expensive and has a high CTE;
- AlN has a low dielectric, and high thermal conductivity and a low CTE, but is expensive;

- Glass, including fused silica, is inexpensive, has a low dielectric, a low CTE but a very low thermal conductivity, but is still favored, because of low cost and low CTE and low dielectric;
- We believe that silicon has not been used for two key reasons: one being that milled silicon has till today not been available because of the danger of milling silicon to sizes in the micron to submicron range, and secondly, that unless it is pure it may be too electrically conductive to use as a filler for EMC bodies.

Therefore as an example we propose the use of silicon milled according to US Patent 6,638,491 28 October 2010 Carberry. In such a case the use of this technology is helpful in that it provides for a safe cost effective way to mill silicon. Furthermore, double milling trials, using two different size milling media in sequence have provided very attractive tri and bi modal particle size distributions helpful in achieving high solids loading in epoxy.

The advantages of silicon include the very high thermal conductivity, the low CTE, relatively low cost:

- Thermal Conductivity at 145 W/M*k is:

    o In the ball park of commercially available AlN,
    o Slightly inferior to commercially available BeO,
    o An order of magnitude better than Al2O3; and
    o Two orders of magnitude better than Fused Silica.

- Silicon's CTE at 2.8-3.2 is:

    o Only Fused Silica is lower at .5 PPM;
    o Only AlN is close, at 4.5;
    o BeO and Al2O3 are both much higher at 8;
    o But at 3 PPM it should with proper solids loading be able to modify the epoxy resin, which is 15-100 PPM, to a low value of around 5 PPM must as fused silica does.

- The Fracture Toughness (K1C) of silicon is about 1,which is similar to fused silica. This means one can expect the same milling fracture behavior, morphology, particle size and particle size distribution as with fused silica;
- In addition one can expect the same or superior chemical reaction between the surface of the silicon with silanes used in the process of making epoxy molding compounds as is experienced with silica.
- In terms of cost, only fused silica will be less costly, but silicon will be in the same order of magnitude, while all others are going to be very expensive.

Therefore the remaining challenge is the electrical properties. Given the great benefit silicon offers in all other regards, it is worthy therefore to solve this problem so to enable packaging with much greater thermal dissipation capability. These we propose to modify in one or both of several approaches:

a) In our milling process we propose in one version of our invention to use a Passivation agent, for example ethyl silicate, as a Passivation agent to deposit a thin layer of glass on the surfaces of the powders as we mill them, creating an attractive surface dielectric property on these surfaces;

b) We also propose to use the epoxy itself in a modified form so to passivate the surface of the powders;

c) Finally we propose to passivate the surfaces of the semiconductor structures themselves before submitting to epoxy molding compound.

CONCLUSIONS

Current epoxy molding compounds are filled with quartz or fused silica having a thermal conductivity of about 1.45 watts, and are filled by volume to about 50% filler, 50% resin. The thermal conductivity of the resin is about .3 watts, so that the resulting epoxy molding compounds have thermal conductivity of about .7-.9 watts. Silicon has nearly ideal properties with regard to the CTE (about 3 ppm) and thermal conductivity (about 150 watts) and cost, and should allow us to make epoxy molding compounds with an order of magnitude or more increase in thermal conductivity.

We would note that in making the silicon filler for these epoxy molding compounds it is important to avoid fractions less than perhaps 20 microns, which would have grain boundary interface areas of about 500 nanometers, the threshold for the phenomena of phonon blocking which is understood the reduce thermal conductivity dramatically.

REFERENCES

[1] Xingzhe Wang et al, Effect of Grain Sizes and Shapes on Phonon Thermal Conductivity of Bulk Thermo Electric Materials published in the *American Institute of Physics*

[2] John B. Butt "Thermal Conductivity of Porous Catalysts" *A.I.CH.E Journal*

[3] Fang and Pilon "Scaling laws for thermal conductivity of crystalline nanoporous silicon based on molecular dynamics simulations" *Journal of Applied Physics*

[4] A. Minnich and G. Chin, *Appl. Phys. Lett.*, 91, 073105 (2007)

[5] M.S. Jeng, R.G. Yang, D. Song and G. Chen, *ASME J. Heat Transfer*, 1 30, 042410 (2008)

[6] K.M. Katika and L. Pilon, *J. Appl. Phys.*, 103, 114308, (2008)

[7] Lee, Galli and Grossman "Nanoporous Si as an Efficient Thermoelectric Material" *Nano Lett.*, 23 October 2008 in the American Chemical Society

[8] Sabah et al. of JPL "Nanostructured Bulk Silicon as an Effective Thermoelectric Material" in *Advanced Functional Materials,* 2009, 19, 2445-2452

[9] Ci et al. [9] "Thermoelectric Properties of Silicon Microchannel Plates Structures" in the *Journal of Physics Conference* series 276 (2011) 012043 and Ci et al "Novel thermoelectric materials based on oron-doped silicon microchannel plates in *Materials Letters* 65 (2011) 1618-1620

[10] Pengliang et al. [10] "Novel Thermoelectric materials based on boron-doped silicon microchannel plates" published in *Materials Letters,* 65 (2011) 1618-1620

[11] John Carberry US Patent 6,638,491 B2

PROCESSING OF INERT SIC MATRIX WITH TRISO COATED FUEL BY LIQUID PHASE SINTERING

Kazuya Shimoda and Tatsuya Hinoki
Institute of Advanced Energy, Kyoto University, Gokasho Uji, Kyoto 611-0011, Japan

Kurt A Terrani, Lance L Snead and Yutai Katoh
Oak Ridge National Laboratory, P.O. Box 2008, Oak Ridge, TN 37831-6138, USA

ABSTRACT
    SiC is a promising material for nuclear applications especially due to its outstanding properties of excellent high temperature strength, high thermal conductivity and exceptional low-activation against neutron irradiation. It is proposed to replace the current uranium oxide fuel pellets for light water reactors with the ceramic micro-encapsulated uranium compacted within SiC inert matrix. In this study, nanostructured SiC fabricated by liquid phase sintering of SiC nanopowder was proposed as a dense inert SiC matrix for compact-packing of surrogate TRISO fuel particles. Sintering additives like Y_2O_3 and Al_2O_3 were used in addition to SiC nanopowder in the amount of 6 wt%. The powder mixture without the fuel particles was sintered by hot-pressing at 1750-1900 °C under 10 MPa to investigate the influences of sintering temperature on density and compression strength in inert SiC matrix itself. Based on the condition for densfied SiC matrix formation without the fuel particles under as a low sintering temperature as possible, pellets including 40 vol% TRISO particles were sintered at hot-pressing and characterized by microstructural observation and evaluation of thermal conductivity.

INTRODUCTION
    There have been many efforts to apply an inert matrix fuel (IMF) concepts as a way to eliminate the current excess stockpiles of plutonium and minor actinides in thermal reactors and advanced nuclear systems during the last few decades [1-5]. Many materials have been proposed as an inert matrix, e.g. pure metals, alloys, oxides, nitrides and carbides. Silicon carbide (SiC) is a promising candidate material because it satisfies required criteria such as light specific weight, high melting point (approximately 2700 °C), relatively high thermal conductivity, resistance to chemicals, excellent thermo-mechanical properties, relatively low coefficient of thermal expansion and good behaviors under irradiation (inherent low radioactivity, no unacceptable phase changes, low neutron absorption and low irradiation swelling). We are now focusing on fully Ceramic Microencapsulated (FCM) fuels, which consist of Tristructural Isotropic (TRISO) particles embedded in a SiC matrix. TRISO coating system has become the most common technique to fabricate fuels for gas fast reactors as well as the next generation high-temperature gas fast reactors. In the conventional (high-temperature) gas-cooled reactor application, the TRISO particles are dispersed in a graphite matrix producing compacts in the form of pebbles or pellets [6, 7]. TRISO particles are in turn comprised of a spherical fuel kernel that is coated with successive layers of porous carbon (buffer layer), a dense inner pyrocarbon (IPyC), SiC, and an outer pyrocarbon (OPyC) layer, respectively. Under the FCM fuel concept, the graphite matrix is replaced with a SiC matrix to offer the following potential advantages, such as improved irradiation stability, effective barrier to fission product release and environmental stability under operating [8]. However, a liquid phase sintering (LPS) process using sintering additives under an applied pressure is adapted for the densification of inert SiC matrix since SiC is difficult to

densify without sintering additives because of its high degree of covalent bonding. The LPS process decreases the sintering temperature of SiC through the formation of liquid phase by 200-400 °C, in comparison to the conventional solid state (pressureless) sintering process using the addition of boron and carbon as sintering additives which requires very high temperature (> 2200 °C) for sintering of SiC. The LPS SiC has been successfully developed with Y_2O_3 or other rare earth oxides and Al_2O_3 or AlN [9-12]. Lowing of the sintering temperature is of great importance when composites' materials, including particulate-, whisker- and fiber-reinforced SiC composites, in this case TRISO fuel particles, are to be added. In this paper, we report on recent activity to fabricate a dense inert SiC matrix with TRISO particles by LPS of SiC nanopowder. In particular, this paper starts to provide the detailed information about SiC nanopowder as a starting material, and then the influences of sintering temperature on microstructure and the properties of inert SiC matrix without TRISO particles. Based on the results without TRISO particles, inert SiC matrix with TRISO particles were prepared and its microstructure and thermal conductivity were analyzed in order to understand the reactions during hot-pressing, to investigate the compatibility of TRISO particles with LPS SiC matrix and to compare the case without TRISO particles.

EXPERIMENTAL PROCEDURE

SiC nanopowder (CEA-Saclay, France, mean particle size <50 nm, 99 % up pure) manufactured by laser pyrolysis method was employed as a starting material in this study. The morphology and microstructure of the SiC nanopowder were observed using field emission scanning electron microscopy (FE-SEM), transmission electron microscopy (TEM) and high resolution TEM. The micro-chemical structure on the nanopowder surface was analyzed by Energy Dispersive X-ray Spectroscopy (EDXS) and X-ray Photoelectron Spectroscopy (XPS, PHI Quantum 2000). Al_2O_3 (Kojundo Chemical Laboratory Co. Ltd. Japan, mean particle size of 0.3μm, 99.99 % pure) and Y_2O_3 (Kojundo Chemical Laboratory Co. Ltd., Japan, mean particle size of 1.0 μm, 99.99 % pure) were used as sintering additives. Al_2O_3 and Y_2O_3 particles (Al_2O_3:Y_2O_3 = 60:40) were added at a relatively small amount of 6 wt%. The SiC nanopowder was ball-milled with the additives in isopropanol for 5 h to form SiC slurry. The mixed slurry was dried, and pressed to form a pellet shape with a diameter of 8mm and a thickness of ~6 mm in dimension. The 3-4 pellets prepared were sintered by hot-pressing in Ar atmosphere under a pressure of 20 MPa for 1h. Sintering temperature was varied from 1750 °C to 1900 °C. The sintered sample was a diameter of 8mm and a thickness of ~6mm in dimension and then mechanically sliced into a thickness of 2mm. Density of the sintered body was determined by the Archimedes principle, using distilled water as the immersion medium. The theoretical density of the sintered samples was calculated by following the ratio of a mixture of SiC powder (3.21 g/cm^3) and sintering additives (4.00 g/cm^3 for Al_2O_3 and 5.03 g/cm^3 for Y_2O_3). The disk compression test was carried out using the number of 3 sliced specimens, with crosshead speed of 0.1 mm/min at room-temperature in an INSTRON 8861 test machine. Compression strength (σ) was calculated by the following equation (1) [13].

$$\sigma = 2P/(\pi Dt) \tag{1}$$

where P is load at given point in test, D is specimen diameter, t is specimen thickness. The thermal diffusivity and the heat capacity in the perpendicular face to the hot-pressing direction were measured using the sliced specimens by a laser flash technique at a temperature region of 25-1200 °C in vacuum using a TC7000 test machine (ULVAC-RIKO, Inc. Japan). The thermal diffusivity (α) was determined from by $t_{1/2}$ method using the following equation (2) [14]:

$$\alpha = 1.38L^2/\pi^2 t_{1/2} \tag{2}$$

where $t_{1/2}$ is the time required to reach half of the total temperature rise on the near surface of the specimen and L is the specimen thickness. The thermal conductivity (K) in the direction through the thickness was calculated from the following thermal diffusivity, the specific heat and the density of sintered samples:

$$K = \alpha \rho C v \tag{3}$$

where α is the thermal diffusivity, ρ is the density and C_v is the specific heat. TRISO particles with spheroidal ZrO_2 particles for kernel were used as surrogate fuels in this study. The 40 vol% TRISO particles-the powder mixture including SiC nanopowder and 6 wt% sintering additives was pressed to form a pellet shape with a diameter of 8mm and a thickness of ~6 mm. The 3-4 pellets were sintered by hot-pressing at 1850 °C in Ar atmosphere for 1h. The obtained thermal result was compared with that of inert SiC matrix without TRISO particles (monolithic SiC) under the same hot-pressing condition.

RESULTS AND DISSCUSSIONS

Densification process

Fig. 1 shows morphologies and microstructures observed FE-SEM, TEM and high resolution TEM of SiC nanopowder used in this study. The average particle size was 40 nm. All particles were controllable less than 100 nm with a very sharp peak, which indicated narrow-distribution. Surface impurity was not detectable by FE-SEM, TEM and high resolution TEM. EDXS analysis identified C and Si without O on the selected point near powder surface. This suggested that this powder contains little oxide-based impurity phase, in particular oxide amorphous phase. Surface chemical structure of SiC particles usually exists as the condition that SiC particle is covered with free carbon, and oxide (amorphous) phase like silica (SiO_2) or silicon oxycarbide (SiC_xO_y) [15, 16]. Detailed chemical characterization for impurity phase was performed by XPS. The present authors reported that XPS was an effective technique to investigate impurity phase on the surface of SiC nanopowder by chemical bond analysis [16]. Fig. 2 shows narrow scanning results (the spectra for O_{1s} (E_b = 542–525 eV), C_{1s} (E_b = 292–278 eV) and Si_{2P} (E_b = 107–97 eV)

Fig.1 FE-SEM, TEM and high resolution (HR) TEM images of SiC nanopowder used in this study.

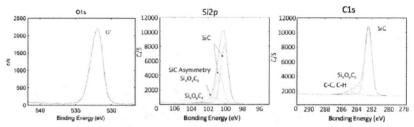

Fig. 2 Narrow scanning results (the spectra for O_{1s} (E_b = 542–525 eV), C_{1s} (E_b = 292–278 eV) and Si_{2P} (E_b = 107–97 eV) signals) of SiC nanopowder using XPS.

signals) of SiC nanopowder using XPS. The nanopowder used in this study was composed predominantly of Si and C with much lower levels of O. The low levels of O detected are consistent with that expected for several structural models, such as C–O–C or Si–O–Si bridges, or an insertion of O atom into back-bond of Si (Si–O–C). However, it was difficult to identify the origin of the O′ peak clearly in the present data. C is present predominantly as SiC with much lower levels of hydrocarbon (likely adventitious in nature), and a signal at 283.7 eV which is most likely due to SiC_xO_y. Si is present predominantly at a binding energy consistent with other references for SiC. Much lower levels of signals at 101.4 and 102.9 eV are also present. The latter is most likely due to SiC_xO_y while the former is due to either SiC_xO_y and/or asymmetry from the SiC signal. Fig. 3 shows (a) relative density and (b) compression strength of inert SiC matrix without TRISO particles (monolithic SiC) as a function of sintering temperature. It is very important to investigate the densification of pellet-shaped inert SiC matrix itself without TRISO particles before processing of inert SiC matrix with TRISO particles. For composites' fabrication by hot-pressing, potential particulate/whisker/fiber degradation caused by high temperature and pressure needs to be considered [17-19]. If the SiC matrix could obtain sufficient densification, the sintering temperature and applied pressure should be as low as possible. We reported that SiC nanopowder as a starting material enhanced densification through LPS below 1900 °C, compared with SiC sub-micron powder due to the promoted boundary- and volume-diffusion from boundary by the increased interparticle contact points derived from finer particles [20].

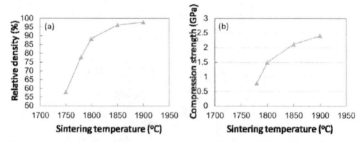

Fig. 3 (a) relative density and (b) compression strength of inert SiC matrix without TRISO particles (monolithic SiC) as a function of sintering temperature.

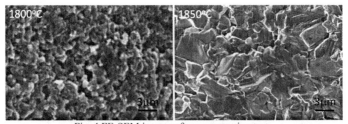

Fig. 4 FE-SEM images after compression test.

The densification was dramatically promoted at the temperature of 1750-1800 °C and saturated above 1800 °C, which had the close relationship with eutectic phase point of binary Al_2O_3-Y_2O_3 system (T=1760 °C). The mechanical property was a similar behavior to the densification. High densification (> 95 %) with high compression strength (> 2 GPa) can be obtained from 1850 °C. This can also be demonstrated by observing the microstructural evolution as shown in Fig. 4. The homogenous grain growth of fine SiC particles and rigid grain boundary structure developed at 1850 °C, resulting in the progress of densification and the decrease of pores on triple points.

Fig. 5 shows digital images of a pellet with 40 vol% TRISO particles, sintered at 1850 °C under 20 MPa. We have successfully developed near-net (pellet) shape technique with a diameter of 8 mm and a thickness of 6mm for inert SiC matrix fuels in this study. The bulk density was ~2.7 g/cm³, which corresponded to be about 90 % of theoretical density. TRISO particles were well distributed in the SiC matrix and significant damages of TRISO particles like cracks and fractures were not detectable at both center and edge parts. However, some pores (open porosity: ~ 4 %) were observed in inert SiC matrix.

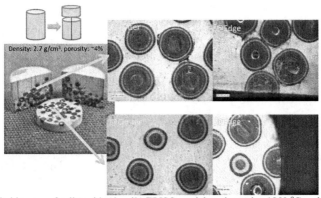

Fig. 5 Digital images of pellet with 40 vol% TRISO particles, sintered at 1850 °C under 20 MPa.

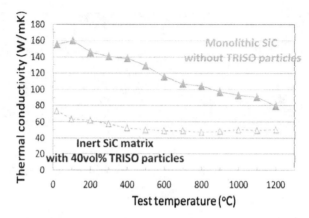

Fig. 6 Thermal conductivities of inert SiC matrix with or without 40 vol% TRISO particles.

Fig. 6 shows thermal conductivities of the inert SiC matrix with and without TRISO particles as a function of test temperatures. Thermal conductivities of inert SiC matrix without TRISO were 151 W/m*K at 25 °C and over 90 w/m*K at 1000 °C, which was quite higher than conventional LPS SiC with Al_2O_3-Y_2O_3 additives using sub-micron powder (20-40 W/m*K at 1000 °C) [21]. This might be due to the high densification with very low porosity, relatively large grains and less additive content for densification. Whereas, thermal conductivities of inert SiC matrix with 40 vol% TRISO particles were 70 W/m*K at 25 °C and near 50 W/m*K at 1000 °C. The thermal conductivity value at 25 °C was more than 75% of the roughly estimated thermal conductivity value by the simple rule of mixtures ($K_{TRISO(ZrO2\ kernel)}$: ~4 W/m*K at 25 °C). This experimental value was a little lower than the roughly estimated value mainly due to the dispersed pores (~4 %) in the SiC matrix. However, the inert SiC matrix with 40 vol% TRISO particles developed in this study possessed thermal conductivity values much higher than others previously reported as IMFs up to high temperatures, owing to the high thermal conductivities of inert SiC matrix without TRISO particles itself [21]. It has been reported that the thermal conductivity of SiC irradiated in a fast neutron flux decreases significantly [22].

SUMMARY
We addressed inert SiC matrix to replace the graphite matrix in the TRISO fuel system. In particular, the LPS process using SiC nanopowder was employed for dense inert SiC matrix formation. Main highlights in this study can be drawn:
(1) SiC nanopowder used in this study was characterized in details by micro-chemical and -structural analyses. The average particle size was 40 nm and all particles were controllable less than 100 nm. The powder surface was quite high purity with very low content of oxygen.
(2) The densification of inert SiC matrix was dramatically promoted at the temperature of 1750-1800 °C and saturated above 1800 °C. The mechanical property was a similar behavior to the densification. High densification (> 95 %) with high compression strength (> 2 GPa) can be obtained from 1850 °C.

(3) We have successfully developed near-net (pellet) shape technique with a diameter of 8 mm and a thickness of 6mm for inert SiC matrix fuels, whose bulk density was ~2.7 g/cm^3 (90 % achievement of theoretical density). TRISO particles were well distributed in SiC matrix and took no significant damages like cracks and fractures. Thermal conductivity of 40 vol% TRISO particles exhibited excellent values up to 1200 °C , owing to the high thermal conductivity values of inert SiC matrix without TRISO particles itself.

REFERENCES
[1] "Ceramic Matrix Composites -Microstructure, Properties and Applications-," edited by I. M. Low. Woodhead Publishing, Abington, England, 2006.

[2] B. Riccardi, L. Giancarli, A. Hasegawa, Y. Katoh, A. Kohyama, R.H. Jones and L.L. Snead, "Issues and advances in SiC$_f$/SiC composites development for fusion reactors," *Journal of Nuclear Materials.*, **329–333**, 56–65 (2004).

[3] R. Naslain, "Design, Preparation and Properties of Non-Oxide CMCs for Application in Engines and Nuclear Reactors: An Overview," *Composites Science and Technology.*, **64**, 155-170 (2004).

[4] A. Kohyama, S. M. Dong and Y. Katoh, "Development of SiC/SiC Composites by Nano-Infiltration and Transient Eutectic (NITE) Process," *Ceramic Engineering and Science Proceedings.*, 311-318 (2000).

[5] S. M. Dong, Y. Katoh and A. Kohyama, "Processing Optimization and Mechanical Evaluation of Hot Pressed 2D Tyranno-SA/SiC Composites," *Journal of the European Ceramic Society.*, **23**, 1223-1231 (2003).

[6] J. A. DiCarlo, "Microstructural Optimization of High Temperature SiC/SiC Composites," Proceedings of the 5th International Conference on High Temperature Ceramic Matrix Composites (HTCMC5), p. 187-192, The American Ceramics Society, Ohio (2004).

[7] Y. Katoh, A. Kohyama and T. Nozawa, "SiC/SiC Composites through Transient Eutectic-phase Route for Fusion Applications," 11th International Conference on Fusion Reactor Materials (ICFRM-11) Presented at, Kyoto, Japan (2003).

[8] T. Ishikawa, Y. Kohtoku, K. Kumagawa, T. Yamamura and T. Nagasawa, "High-Strength Alkali-Resistant Sintered SiC fiber Stable to 2200°C," *Nature.*, **391**, 773-775 (1998).

[9] K. Shimoda, J.S. Park, T. Hinoki and A. Kohyama, "Densification Mechanism and Microstructural Evolution of SiC Matrix in NITE Process", *Ceramic Engineering and Science Proceedings.*, **27** [5], 19-27 (2006).

[10] F. Rebillat, J. Lamon, R. Naslain, E. L. Curzio, M. K. Feber and T. M. Besmann, "Properties of Multilayered Interphases in SiC/SiC Chemical-Vapor-Infiltrated Composites with 'Weak' and 'Strong' Interfaces," *Journal of the American Ceramics Society.*, **81** [9], 2315-2326 (1998).

[11] F. Rebillat, J. Lamon and A. Guette, "The Concept of a Strong Interface Applied to SiC/SiC Composites with a BN Interphase," *Acta Materialia.*, **48**, 4609-46018 (2000).

SiC-COATED HTR FUEL PARTICLE PERFORMANCE

Michael J. Kania, Averill Park, NY, USA
Heinz Nabielek, Düren, NRW, Germany
Karl Verfondern, Forschungszentrum Jülich, Germany

ABSTRACT

The German HTR (High Temperature Gas-Cooled Reactor) fuel program successfully developed, licensed and manufactured many thousands of spherical fuel elements that were used to power the experimental AVR reactor and the prototype THTR[1] reactor. After completing fuel development for AVR and THTR with BISO (bi-isotropic) coated particles, the German program continued toward a new development program utilizing TRISO (tri-isotropic) coated particles. Advanced HTR concepts where spherical fuel elements with TRISO-coated particles are applicable were: process heat application; direct-cycle electricity production with a gas turbine in the primary circuit; and conventional steam generation.

In TRISO coated particle development, the combination of low-temperature-isotropic (LTI) inner and outer PyC layers surrounding a strong, stable SiC layer greatly reduced heavy metal contamination levels and defective particle fractions in production fuel elements. In addition, the TRISO coated particle provided improved mechanical strength and a higher degree of solid fission product retention, not known previously with BISO coatings.

The improved performance of the HEU $(Th,U)O_2$ and LEU UO_2 TRISO fuel systems was successfully demonstrated in all three areas: manufacturing, irradiation performance under normal operating conditions, and accident condition performance. Good irradiation performance for these fuel systems was demonstrated under normal operating conditions to 12% FIMA and in accident conditions not exceeding 1600°C.

INTRODUCTION

The modern TRISO particle development was performed with both high-enriched uranium (HEU) and low enriched uranium (LEU) dioxide fuels manufactured in the period 1978-1988 and irradiation tested from1978 to 2010. Large scale manufacturing campaigns produced high quality spherical elements for AVR Reloads 15, 19, 20, and 21 (Table 1). In addition to real-time testing in the AVR, fuel elements from these campaigns were irradiation tested in European Material Test Reactors. Irradiated elements were then subjected to post-irradiation examinations (PIE) including accident simulation tests. Together, the AVR and the MTR tests covered a wide range of HTR applications[2],[3],[4],[5] suitable for

- 750°C coolant exit power generation with a steam turbine,
- 850°C coolant exit power generation with a direct cycle gas turbine,
- 950°C coolant exit for process heat applications like hydrogen generation.

Table 1: High-quality AVR Reload fuel manufacturing campaigns (1978 – 1988).

AVR Reload Number	15	20	19	21	21/2
Element Population	6,083	11,850	24,611	20,350	8,740
Element Designation	GO 2		GLE 3	GLE 4	
Fuel Particle Type	HEU (Th,U)O$_2$ TRISO		LEU UO$_2$ TRISO		

A historical overview of the irradiation tests conducted within the German Fuel Development Program and the more recent European Union Fuel Development Program is given in Table 2. From 1977 to 1981, a significant number of irradiation tests were executed with the HEU (Th,U)O$_2$ TRISO fuel system in support of the Process Heat and Direct Cycle HTR concept development. In the period 1981 to 2011, additional irradiation tests were executed in support of LEU UO$_2$ TRISO fuel development for the Modular Pebble Bed Reactor. (In Table 2, the designations DR, HFR, FRJ2, BR2, AVR are acronyms for the reactors; and the letters P, M, S designate particle, compact and spherical fuel element tests).

HEU (TH,U)O$_2$ AND LEU UO$_2$ TRISO PARTICLE DESIGNS

High quality TRISO coated fuel particles were produced in full-size production facilities by NUKEM in Germany. The dimensions of the TRISO particle design are shown in **Table 3**.

The TRISO particles were fabricated into either: special irradiation test specimens and subsequently tested in accelerated irradiation MTR tests; or 60-mm diameter reference spherical elements as part of the five large AVR TRISO fuel element reload campaigns.

The design functions of the ceramic coating layers of the TRISO coated fuel particle are described in Table 4. There is little difference between the TRISO coatings applied to HEU (Th,U)O$_2$ LTI TRISO and LEU UO$_2$ LTI TRISO particle designs (LTI = Low Temperature Isotropic pyrocarbon). All four layers are deposited in an uninterrupted sequential chemical vapor deposition process in the same fluidized-bed coating furnace (**Table 5**). The conditions under which each layer deposition takes place are important as they determine its material properties. Process parameters such as time, temperature, pressure, gas composition and gas ratios all play important roles in fixing these properties.

Key material property requirements for good irradiation performance in the dense isotropic PyC layers and near-theoretical dense SiC layer are:

the PyC layer:
- must be impermeable
- have an isotropic texture
- deposited at low enough temperature to avoid heavy metal contamination

the SiC layer:
- β-SiC with a cubic structure of type 3C
- a density > 3.19 g/cm^3
- equiaxed microstructure with fine grains and few flaws
- PyC-SiC interfaces must be of sufficient strength

Table 2: Historical overview of irradiation tests conducted within the German and the more recent European Union Fuel Development Programs. The HEU/Th TRISO program is marked by black borders, the LEU TRISO program is in green.

R&D program	Old LEU	HEU Program for Process Heat and Gas Turbine Applications			LEU Program
	1972-76	1977-1981			1982-2010
Coated particle	UO_2 TRISO	Variant 1	Variant 2	Variant 3	UO_2 TRISO
	UO_2 BISO	$(Th,U)O_2$ BISO	$(Th,U)O_2$ TRISO	UCO TRISO+ ThO_2 TRISO	
Test goal					
Particle performance	HFR-M5	BR2-P24	BR2-P25	BR2-P23	HFR-P4
	DR-S6				SL-P1
Fission product transport in intact particles	DR-S4	FRJ2-P22	FRJ2-P23	FRJ2-P24	FRJ2-P27
Release from kernel	-	FRJ2-P25	FRJ2-P25	FRJ2-P25	FRJ2-P28
Chemical effects	FRJ2-P16	-	-	HFR-P3	HFR-P5
Fuel element tests					
Fuel element performance	DR-K5	HFR-K1	R2-K12 R2-K13	-	HFR-K3
FE fission product transport	-	-	FRJ2-K11	FRJ2-K10	FRJ2-K13 FRJ2-K15
Large-scale demonstration	AVR 6	AVR 14 AVR 18	AVR 15 AVR 20	AVR 13	AVR 19 AVR 21
Proof tests					HFR-K5 HFR-K6
High burnup tests					FRJ2-K15 HFR-EU1
High temperature tests for hydrogen production etc.					HFR-EU1bis

Table 3: Nominal dimensions of the HEU (Th,U)O$_2$ LTI TRISO particle design.
The LEU UO$_2$ LTI TRISO particles are much the same, only the buffer layer is thicker.

TRISO Particle Component	Dimensions (μm)
oxide kernel diameter	500
Buffer Layer Thickness	90 (95)
iPyC Layer Thickness	40
SiC Layer Thickness	35
oPyC Layer Thickness	40

Table 4: Design functions of ceramic coating layers in a TRISO coated fuel particle.

TRISO Particle Coating Layer	Design Function
Buffer (50% dense PyC)	• Provides void volume for gaseous fission products and carbon-oxygen reaction products (CO, CO$_2$) released from fuel kernel • Accommodate fuel kernel swelling • Protects PyC and SiC layers from fission product recoil
inner PyC (density ≥1.85 g cm^{-3})	• Diffusion barrier to fission products, retain gaseous fission products • Impermeable layer prevents Cl$_2$ from reaching kernel during SiC deposition, and prevents CO from interacting with SiC during irradiation • Provides mechanical substrate for deposition of SiC layer
SiC (near theoretical density of 3.21 g cm^{-3})	• Primary barrier to fission products, retains all gaseous and solid fission products at normal operating temperatures (<1250°C) • Load bearing layer for particle
outer PyC (density > 1.85 g cm^{-3})	• Creates compressive stress on SiC during irradiation due to PyC shrinkage • Retain gaseous fission products • Provides bonding layer with carbonaceous fuel element matrix

Table 5: Typical processing parameters for the deposition of the LTI-TRISO coating.

Coating Layer	Decomposition Gas	Carrier Gas	Deposition Temperature (°C)	Deposition Rate (μm/min)
Low density carbon	C_2H_2	Argon	1250	10
Inner dense isotropic PyC	Mixture of C_3H_6 and C_2H_2	Argon	1300	5
Isotropic SiC	CH_3SiCl_3	Hydrogen	1500	0.2
Outer dense isotropic PyC	Mixture of C_3H_6 and C_2H_2	Argon	1300	5

The TRISO coated particles exhibit PyC coatings with typically high densities (close to PyC theoretical density of 2.0 g/cm^3), low anisotropy and thicknesses close to the nominal values listed in Table 3. Similarly, the SiC layers in each particle batch have thicknesses close to the nominal value of 35 μm, densities ≥ 3.19 g/cm^3, and defect fractions, as measured by the Burn Leach Test, typically in the low 10^{-6} range.

The final production steps in TRISO particle production are sieving to remove any under- and over-sized particles, followed by sorting to remove any odd-shaped particles. Sorting is performed on a vibrating table slightly inclined to allow spherical particles to roll down-hill following a parabolic path while odd-shaped particles are vibration transported along a perpendicular direction and recollected for recycling.

SPHERICAL FUEL ELEMENT PRODUCTION

The fabrication process for the HTR spherical elements is described in the pictorial flowchart provided in Figure 1. The steps in spherical fuel element processing are:

- resinated graphitic matrix powder preparation
- overcoating the TRISO particles
- pre-molding the fuel zone
- high-pressure isostatic cold pressing of the complete fuel element
- machining
- carbonization at 800°C, and
- final heat treatment at 1900-1950°C

This same or nearly similar spherical fuel element fabrication process has been used in the past in Russia and South Africa, and is currently being employed in China for fabrication of reference 60 mm-diameter fuel elements for the HTR-PM concept.

The essential characterization technique for quality assurance of high-quality spherical fuel elements is the Burn-Leach Test for HTR fuel. Here, the carbonaceous material of the sample to be measured (loose coated particles, spherical element, fuel compact or coupon) is burnt in a combustion chamber at around 800°C in air down to the SiC layer of the coated particles. Burning is continued until the weight remains constant (~90 hours for a spherical fuel element). The residual of ash and particles is treated with a nitric acid solution at 100°C and the amount of dissolved uranium and thorium quantitatively analyzed. Since the SiC layer is corrosion resistant, the heavy metal found in the solution includes the natural U/Th-content from the matrix material and the U/Th content from those coated particles with defective SiC layers. Also, particles with an incomplete coating will be identified. Burn-Leach Test results are presented as the ratio of the measured free uranium to the total uranium contained in the spherical element, U_{free}/U_{total}. The detection limit is typically at 1×10^{-6} to

3×10^{-6}; this uncertainty is much lower than the heavy metal inventory in a single defective coated particle which is 600 to 700 µg.

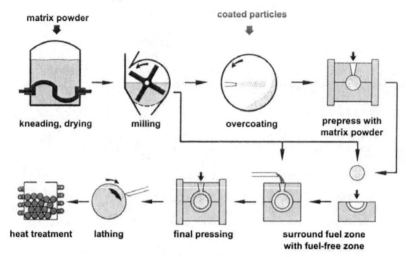

matrix powder　　　　　　　　**coated particles**

kneading, drying　　**milling**　　**overcoating**　　**prepress with matrix powder**

heat treatment　**lathing**　**final pressing**　**surround fuel zone with fuel-free zone**

Figure 1: Pictorial flowchart of the German fabrication method for spherical fuel elements[6,7].

The burn leach results for the five German produced AVR fuel campaigns of Table 1 are detailed in Table 6, Figure 2a. The comparison of German, South African and Chinese LEU UO_2 burn-leach results presented in Figure 2b, are well suited for a modern, inherently safe, small modular HTRs. The extremely low levels of AVR 20 with the equivalent of zero defects and AVR 21/2 with a defect fraction 8×10^{-6} are of the highest quality for any HTR fuel production campaign in their respective development lines.

Table 6: Burn-leach statistics for modern oxide TRISO fuels fabricated in AVR Reloads over the period 1977-1985 (n = number of defective particles; N = population of particles tested).

	Reload	Year of starting manufacture	No. particles per sphere	No spheres tested	No. particles tested N	No. defect particles n	Expected defect fraction =n/N	Upper 95% limit defect fraction*
HEU	AVR 15	1978	10,480	16	167,680	5	2.98E-05	6.27E-05
	AVR 20	1983	10,660	30	319,800	0	0	9.37E-06
LEU	AVR 19	1981	16,400	70	1,148,000	56	4.88E-05	6.09E-05
	AVR 21	1983	9,560	55	525,800	24	4.56E-05	6.42E-05
	AVR 21/2	1985	9,560	40	382,400	3	7.85E-06	2.03E-05

*The upper 95% limit can be obtained from the Excel function BetaInv(0.95, n+1, N+1-n).

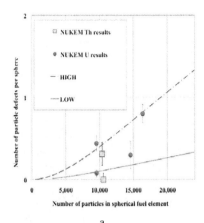

a.

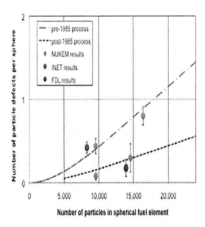

b.

Figure 2: Equivalent TRISO coated fuel particle defects per spherical element as a function of particles/element. Results based on burn-leach tests on statistically significant samples sizes: a) German (Th,U)O$_2$ TRISO and German UO$_2$ TRISO results; and b) comparison of German, South African and Chinese (HTR-10 Production) UO$_2$ TRISO results.

The Burn-Leach data indicate a step-change improvement in spherical fuel element production with TRISO fuel particles in the time period 1983-85, both for HEU (Th,U)O$_2$ and LEU UO$_2$ fuels. In this period, the tabling of kernels, particles and overcoated particles was improved and the manual overcoating process was replaced by an automated process.

HEU (Th,U)O$_2$ IRRADIATION BEHAVIOR (NORMAL OPERATING CONDITIONS)
Within a fuel development program, irradiation testing of prototypical fuel specimens under the anticipated operating conditions of design HTR concepts has been an accepted methodology by which fuel performance data on candidate fuel systems are evaluated. For the HEU (Th,U)O$_2$ TRISO fuel system, a series of fuel irradiation tests were planned and executed on high quality fuels manufactured to well-defined specifications. The irradiation testing included accelerated tests in MTRs together with real-time testing in the AVR. The timeline for these tests is shown in Figure 3. The two large-scale AVR fuel element reload campaigns, AVR 15 and AVR 20, contained HEU (Th,U)O$_2$ TRISO fuel element and were produced in large numbers in production scale facilities by NUKEM. The fuel elements in these reload campaigns were designated AVR GO2.
In a similar manner, an extensive program to qualify high-quality LEU UO$_2$ TRISO fuels was carried out in the German Fuel Development Program from 1982 to 1995. Both accelerated MTR and real-time AVR irradiation testing was planned and executed (Figure 4). The most recent irradiation tests, designated HFR-EU1bis and HFR-EU1, were executed in the 2006-2010 period and employed AVR GLE4 fuel elements from AVR Reload 21/2.
The European MTRs utilized for the accelerated testing of HEU (Th,U)O$_2$ TRISO coated particles typically have higher thermal, epithermal and fast neutron fluxes than an actual HTR. In these environments, fissile fuel burnups and accumulated neutron fluences can be achieved in one to two years, as compared to a typical four year cycle in an HTR. Thus, the term "accelerated" refers to an

irradiation under higher thermal- and/or fast neutron flux environments for the purpose of speeding up the normal rate of fuel burnup and fast fluence accumulation. It is, however, important that target values for burnup, temperature and fluence are well-covered.

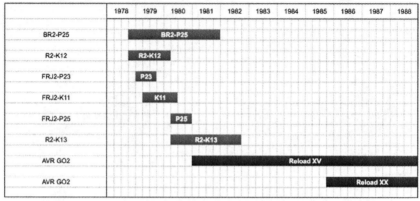

Figure 3: Timeline for HEU (Th,U)O$_2$ LTI TRISO fuel irradiation tests in the period 1978-1988 in MTRs (red) and the AVR (blue).

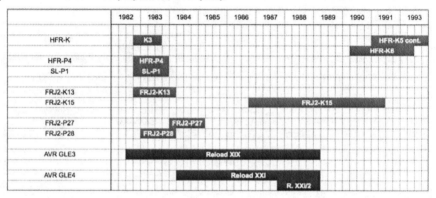

Figure 4: Timeline for with LEU UO$_2$ LTI TRISO fuel irradiation tests starting in 1982 in MTRs (red) and AVR (blue).

Irradiation temperatures are typically maintained within the same operating range as expected in a HTR. This is accomplished by incorporating active temperature control system into the design of the irradiation tests, and together with stepped gas-gaps and precise tolerances on capsule internal components, it is possible to maintain fuel operating temperatures within the acceptable limits. Each independent capsule is swept with a variable mixture of helium and neon purge gas. By varying the composition of this gas mixture, the thermal conductivity of the purge gas located in the control gaps can be adjusted to maintain design temperatures. In the beginning, when the fission rate is high (high heat production), higher concentrations of helium gas (with a high thermal conductivity) are used. As fuel burnup increases, the fission rate in the fuel decreases and higher concentrations of neon gas (with a low thermal conductivity) are employed. In addition to active temperature control, internal thermocouples make it possible to monitor operating temperatures. Additional instrumentation, both active and passive, is generally included in each irradiation test to provide thermal- and fast-flux information and accumulated fast and thermal fluences.

The nominal maximum design operating conditions applicable to the process heat PNP[8] and direct cycle gas turbine HHT[9] concepts are shown below. The seven MTR accelerated irradiation tests containing HEU $(Th,U)O_2$ TRISO fuels were patterned after these operating requirements:

Operating Parameter	Nominal Maximum for PNP and HHT
Fuel Element Central Temperature (°C)	1020
Fuel Burnup (% FIMA)	11
Accumulated Fluence $(x10^{25}$ neutrons/m^2, E>16 fJ)	4.5

In Figure 5, the nominal maximum expected PNP/HHT HTR concepts limit of fast fluence versus temperature is compared with the accumulated fast fluence as a function of irradiation temperature for the six HEU $(Th,U)O_2$ TRISO fueled accelerated MTR tests. These same operational test data are numerically presented in Table 7. Test specimens irradiated in BR2-P25, R2-K12/1 and 2, and R2-K13/1 and 4 well-exceeded the nominal PNP/HHT maximum fast fluence limit and, with the exception of some specimens in R2-K13/4, these fluences were accumulated at operating temperature higher than the PNP/HHT maximum limit. For those specimens in FRJ2-P25/ 2 and FRJ2-P23/1, 2, 3, 4 the accumulated fast neutron fluences were in the mid to lower range of the PNP/HHT concept limit. For FRJ2-P25, these fluences were accumulated in the mid to upper range of the PNP/HHT temperature limit, but for FRJ2-P23, all of the capsules operating temperatures were at or exceeded the PNP/HHT temperature limit, in parts massively so. Also, R2-K12/2 fuel element center temperatures were unusually high, but this did not lead to particle failure even at 12.4% FIMA and a fluence of $6.9x10^{25}m^{-2}$ (E>16 fJ).

For the spherical fuel elements irradiated in FRJ2-K11/3 and 4, the accumulated fluences were at the lowest range (<0.1 x10^{25}) of the PNP/HHT limits. This is due to the location of the fuel elements outside the core of the DIDO reactor where the thermal flux is at a maximum and the fast flux is very low. However, these fluences were accumulated at operating temperatures at the high end or exceeded the PNP/HHT maximum temperature limit.

A comparison of the nominal maximum PNP/HHT HTR concepts fuel burnup and operating temperature limits with those achieved in the six accelerated HEU $(Th,U)O_2$ TRISO fuel MTR irradiation tests is shown in Table 7. Test specimens irradiated in BR2-P25, R2-K12/1,2 and FRJ2-P23/1, 2, 3, 4 achieved burnups above the PNP/HHT limit at operating temperatures well above the PNP/HHT maximum temperature limit. The specimens irradiated in FRJ2-P25/2 (with 1% defects added) achieved burnup above the nominal PNP/HHT maximum, but the operating temperatures ranged from the mid-range to above the PNP/HHT temperature limit. The fuel specimens in FRJ2-K11/3,4 and R2-K13/1 and 4 achieved burnups in the mid to upper range of the PNP/HHT burnup limits with operating temperatures near the upper or well above the PNP/HHT maximum temperature limits. For all of the fuel specimens containing HEU $(Th,U)O_2$ TRISO fuels, the nominal burnup maximum range achieved at EOL was from 8.5% to 15.2 % FIMA over an operating temperature range from 750°C to 1600°C.

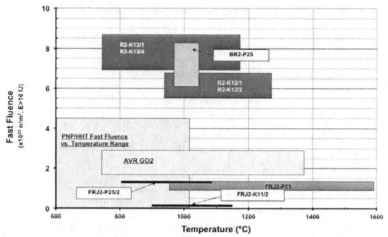

Figure 5: Accumulated fast neutron fluence vs. temperature for HEU (Th,U)O$_2$ TRISO fueled MTR accelerated irradiation tests.

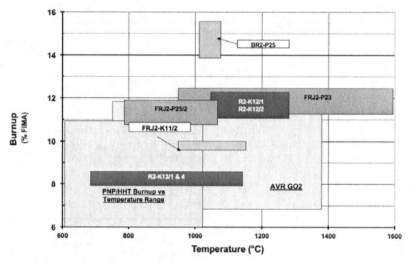

Figure 6: Fuel burnup (% FIMA) vs. temperature for HEU (Th,U)O$_2$ TRISO fueled MTR accelerated irradiation tests.

Table 7: Accelerated MTR irradiation test operating conditions for HEU $(Th,U)O_2$ TRISO fuel.

Irradiation Test/Capsule	Operating Temperature Range (°C)	Burnup Range (% FIMA)	Fluence ($\times 10^{25}$ m^{-2}, E > 16fJ)	^{85m}Kr Release-Rate to Birth-Rate Ratio (R/B)*	
				Beginning of Life (BOL)*	End of Life (EOL)*
BR2-P25*/					
1-12	1010-1070	13.9 – 15.6	6.2 – 8.1	3 x10^{-7}	1 x10^{-6}
R2-K12/					
1	950 - 1100	11.1	5.6	3.9 x 10^{-9}	3.2 x10^{-8}
2	1120 - 1280	12.4	6.9	3.5 x 10^{-9}	3.4 x10^{-8}
FRJ2-P23*/					
1	950 – 1200	11.3	1.1	≤x10^{-7}	1.4 x 10^{-7}
2	1120 – 1200	12.5	1.4	≤x10^{-7}	1.9 x 10^{-7}
3	1330 – 1600	11.9	1.4	≤x10^{-7}	2.3 x 10^{-7}
4	1200 - 1400	12.1	1.1	≤x10^{-7}	2.1 x 10^{-7}
FRJ2-K11/					
3	950 – 1166	9.0	0.062	1.7 x10^{-9}	2.7 x10^{-7}
4	940 - 1162	8.5	0.051		
FRJ2-P25/					
2	850 – 1100	10.7	1.4	9.2 x10^{-7}	1.5 x10^{-5}
R2-K13/					
1	960 – 1170	10.2	8.5	2.2 x10^{-9}	2.1 x10^{-7}
4	750 – 980	9.8	6.8	1.5 x10^{-9}	1.9 x10^{-7}

*The short-lived fission gas ^{88}Kr is shown for the experiments BR2-P25, FRJ2-P23.

LEU UO$_2$ TRISO FUEL

The nominal maximum operating conditions for the general HTR design and the HTR Modul[10] employing modern LEU UO$_2$ TRISO fuel are listed below. All seven of the MTR accelerated irradiation tests and two HTR Module Proof irradiation tests were patterned after these operating requirements.

Operating Parameter	General HTR Design (Phase 1)	HTR Modul[11]
Fuel Element Central Temperature (°C)	1068	870 (Cycled)
Fuel Burnup (% FIMA)	10.1	8.9
Accumulated Fluence ($\times 10^{25}$ neutrons/m^2, E>16 fJ)	2.7	2.1

In Figure 7, the nominal maximum expected general HTR Module Design and Phase 1 UO$_2$ HTR concept limits of fast fluence versus temperature are compared with the accumulated fast fluence as a function of irradiation temperature for the seven LEU UO$_2$ TRISO fueled accelerated MTR tests and the two HTR Modul Proof tests. These same operational test data are numerically presented in Table 8. Test specimens irradiated in HFR-K3 and HFR-P4 well exceeded the maximum fluence and temperature limits.

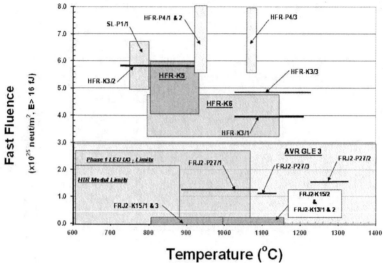

Figure 7: Accumulated fast neutron fluence (x10^{25} neutrons/m^2, E>16 fJ) versus operating temperature for LEU UO$_2$ TRISO coated fuel particles in accelerated irradiation tests conducted in European MTRs prior to 2000 and in the AVR reactor up to December 1988.

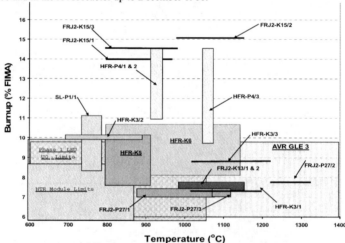

Figure 8: Fuel burnup (% FIMA) versus operating temperature for LEU UO2 TRISO coated fuel particles in accelerated irradiation tests conducted in European MTRs prior to 2000 and in AVR.

Table 8: Operational data during irradiation of modern LEU UO_2 TRISO fuel particles.

Irradiation Test/Capsule	Operating Temperature Range (°C)	Burnup Range (% FIMA)	Fluence (x10^{25} m^{-2}, E > 16 fJ)	^{85m}Kr Release Rate to Birth Rate Ratio (R/B) Beginning-of-Life (BOL)	End-of-Life (EOL)
SL-P1	740 – 790	8.6 – 11.3	5.0 – 6.8	5.8 x10^{-7}	1.2 x10^{-6}
HFR-K3/					
1	1020 – 1200	7.53	4.0	1 x 10^{-9}	2 x10^{-7}
2,3	700 – 900	10.1-10.2	5.8 – 5.9	9 x 10^{-10}	1 x10^{-7}
3	1020 - 1220	9	4.9	2 x10^{-9}	3 x10^{-7}
HFR-P4/					
leg 1	915 – 940	11.1 – 14.7	5.5 – 8.0	3.5 x10^{-9}	8 x 10^{-8}
leg 3	1050 - 1075	9.9 – 14.7	5.5 – 8.0	3.6 x10^{-9}	8 x 10^{-9}
FRJ2-K13/					
1-2	985 – 1150	7.5 – 8	<0.2	2 x10^{-9}	1.6 x10^{-8}
3-4	980 – 1150	7.6 – 7.9	<0.2	8 x10^{-10}	7 x10^{-9}
FRJ2-K15/					
1	800 – 970	14.1	<0.2	2.0 x10^{-10}	1.2 x10^{-8}
2	980 – 1150	15.3	<0.2	2.5 x10^{-10}	5 x10^{-9}
3	800 – 990	14.7	<0.2	2.0 x10^{-10}	3 x10^{-9}
FRJ2-P27/					
1	880 – 1080	7.2 – 7.6	1.4	1.0 x10^{-6}	1.6 x10^{-6}
2	1220 – 1320	8.0	1.7	8.6 x10^{-7}	1.0 x10^{-5}
3	1080 - 1130	7.2 – 7.6	1.3	2.0 x10^{-8}	1.2 x10^{-7}
Proof Test HFR-K5/					
1	800 – 923	7.81	4.0	4.8 x10^{-10}	1.6 x10^{-7}
2	800 – 909	10.06	5.8	2.7 x10^{-10}	3.1 x10^{-7}
3	800 – 903	10.30	5.9	"	"
4	800 – 921	9.26	4.9	2.5 x10^{-7}	3.5 x10^{-7}
Proof Test HFR-K6/					
1	800 – 1090	8.34	3.2	5.0 x10^{-10}	1.5 x10^{-7}
2	800 – 1130	10.64	4.6	3.0 x10^{-10}	4.4 x10^{-7}
3	800 – 1140	10.88	4.8	"	"
4	800 – 1130	9.89	4.5	4.5 x10^{-7}	8.5 x10^{-7}

Accelerated irradiation experiments HFR-K5 and HFR-K6 were Proof Tests for the HTR Modul concept. Each test contained four 60-mm diameter spherical fuel elements in three independent swept gas capsules. For HFR-K5 and -K6, a typical HTR Modul reactor temperature history was simulated during irradiation with 17 temperature cycles, corresponding to 17 fuel element passes through the core. For one third of a cycle, the center temperature of the fuel spheres was held at 800°C, and for the remaining two thirds, the center temperature was maintained at 1000°C. In addition, accident temperature transients for the HTR Modul were also simulated. The spheres center temperatures were increased to 1200°C and held for five hours. These temperature transients were performed at BOL, middle-of-life and EOL in both experiments.

Measured [85m]Kr release rates for HFR-K5 and -K6 over the entire irradiation period are shown in Figure 9 for all six capsules in the two tests. At BOL, the lower set of curves indicate no manufacture defective fuel particles present in Capsules 1,2, & 3 for each of HFR-K5 and -K6. Release data at EOL for these same capsules indicate no irradiation-induced failure in any of the six spherical fuel elements in these same capsules. In contrast, the upper two release curves at BOL represent one and two as-manufactured defective particles in HFR-K5 and –K6, respectively. No additional irradiation-induced particle failures were observed in these capsules during operation.

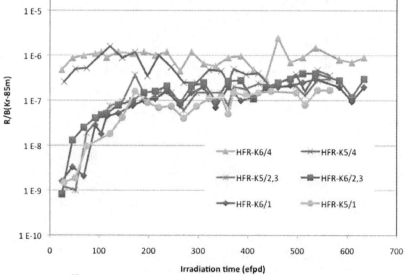

Figure 9: In-reactor [85m]Kr fission gas release-rate to birth- rate (R/B) for irradiation tests HFR-K5 and -K6. The BOL release data indicate one and two manufacturing defects in the fuel elements within HFR-K5/Capsule 4 and HFR-K6/Capsule 4, respectively. The remaining six spherical fuel elements have zero defects at BOL, and developed no induced particle failures during their long-term irradiation.

Figure 7 and Figure 8 compare the accumulated fluence range and burnup range for HFR-K5 and -K6 with the HTR Modul limits. For both Proof Tests, the total fast fluences (E>16 fJ) were well-above the design limit[11] of 2.1×10^{25} n/m² of the HTR Modul concept and were accumulated at temperatures above 800°C. The burnups achieved in all eight fuel elements of the Proof Tests were in the upper portion of the nominal design, ranging from 7.8% to 10.9% FIMA at temperatures from 800°C to 1140°C. The operational test data for HFR-K5 and -K6 are also presented in Table 8.

PREDICTION OF IRRADIATION-INDUCED FAILED PARTICLES IN MTR TESTS

The active temperature monitoring system in each of the MTR accelerated tests made it possible to record temperatures and to measure the release rates of gaseous fission products from the fuel under irradiation. As the helium purge gas exits the individual test capsules, a controlled volume sample is taken and gamma counted to quantitatively determine the quantity of short-lived noble gas fission products as a function of irradiation time. Knowing the sample volume, the purge gas flow rate at time of sampling, and the activity released allows a measure of the fission gas release rate (R_i) directly from the fuel particles. Typically the radioactive krypton and xenon isotopes of interest are ^{85m}Kr, ^{87}Kr, ^{88}Kr, ^{89}Kr and ^{133}Xe, ^{135}Xe, ^{137}Xe and ^{138}Xe. By comparing the measured release-rate of an isotope to the birth-rate (B_i), determined through fuel depletion calculations as a function of time, the ratio $[(R/B)_i]$can provide a direct measure of the steady-state release rate behavior of the fuel. This method of monitoring the fission gas release-rate to birth-rate has become the standard technique of assessing in-reactor fuel performance.

By comparing the EOL ^{85m}Kr or ^{88}Kr R/B values with the R/B of a failed TRISO-coated fuel particle, an estimate of the fraction of failed particles responsible for the fission gas release can be made from:

$$\eta = \frac{\left(R_i\big/B_i\right)_{EOL}}{\left(R_i\big/B_i\right)_f} \qquad\qquad \text{Equation 1}$$

where

η is the derived failure fraction as defined by gas release rates,

$\left(R_i\big/B_i\right)_{EOL}$ is the measured release rate over birth rate for isotope i at end-of-life, and

$\left(R_i\big/B_i\right)_f$ is the predicted release rate over birth rate for isotope i per particle.

Release rates are predicted with the methods and data of IAEA-TECDOC-978 using temperature as the only parameter[12]. Particle failure estimates in MTR irradiated spherical fuel elements based on the above methodology are listed in Table 9.

Table 9: Number of irradiation induced failed particles for spherical fuel elements in MTR tests. The failure has been derived from the comparison with the predicted release rates of short-lived fission gases per particle.

Fuel Type	Irradiation Test/ Elements	Burnup (% FIMA)	Total number of TRISO particles in test	Number of irradiation induced failed particles
HEU/Th	R2-K12/1	11.1	10,830	0
HEU/Th	R2-K12/2	12.4	10,830	0
HEU/Th	FRJ2-K11/3,4	8.5-9.0	20,960	1
HEU/Th	R2-K13/1	10.2	20,050	0
HEU/Th	R2-K13/4	9.8	20,050	2
	sum		*82,720*	*3*
LEU	FRJ2-K13/1 thru 4	7.5-8.0	65,600	0
LEU	FRJ2-K15/1 thru 3	14.1-15.3	28,680	0
LEU	HFR-K3/1 thru 4	7.5-10.2	65,600	0
LEU	HFR-K5/1 thru 4	7.8-10.3	58,400	0
LEU	HFR-K6/1 thru 4	8.3-10.9	58,400	0
	sum		*276,680*	*0*

AVR REAL-TIME IRRADIATION TESTING AND ANALYSIS

The AVR reactor (46 MW_{th}) located in Jülich Germany, operated from 1967 through 1988[13]. During the 21 years of operation it provided invaluable information on spherical fuel element development, fuel particle development with many particle variants (kernel material, enrichments, coating designs) and various HTR fuel cycles. More than 290,000 spherical fuel elements of five different types, containing more than 6×10^9 coated fuel particles, were inserted into its core. The distribution of various fuel element types within the AVR core as a function of operating history are shown in Figure 10[14].

High-quality AVR fuel elements containing the HEU $(Th,U)O_2$ TRISO fuel particle system were fabricated in AVR Fuel Reloads - AVR 15 and AVR 20. These elements were designated the AVR GO2 fuel type and were inserted into the AVR core during an eight year period beginning in 1981 and again in 1985. Three follow-on campaigns which contained high-quality LEU UO_2 LTI TRISO fuel elements were fabricated in AVR Reloads AVR 19, AVR 21 and AVR 21/2. Elements from Reload 19 were designated type GLE3, and those from Reloads 21 and 21/2 were designated GLE4. The modern LEU UO_2 LTI TRISO elements were inserted into the AVR beginning in 1982 with the GLE 3 elements, followed by the GLE 4 elements in 1984 and 1987. Figure 10 graphically displays the distribution of fuel elements inserted into the AVR Core during its operational lifetime. The HEU $(Th,U)O2$ TRISO elements (GO 2) and LEU $UO2$ TRISO elements (GLE 3 and GLE 4) were the last elements inserted into the AVR (designated by blue, green and yellow areas).

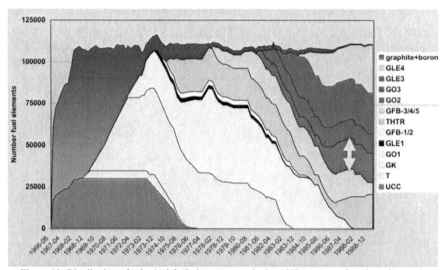

Figure 10: Distribution of spherical fuel element types in the AVR core as a function of operating history. Five distinct types of elements were inserted into the AVR over its operating lifetime (1967 to 1988) comprising ~290,000 elements. High-quality HTR fuel elements containing TRISO coated fuel particles constituted only about half of the core at the end of operations.

At periodic intervals during AVR operation, a number of irradiated elements would be randomly drawn from the core for post-irradiation evaluation and accident condition testing at the FZ-Jülich, Germany. Table 10 is a summary of the average and maximum burnup and accumulated fluence values for the GO2, GLE3 and GLE4 fuel elements withdrawn from the AVR. Post-irradiation evaluations include: fission product inventory measurements (burnup), out-of-reactor gas release measurements, and accident simulation testing.

Table 10: Average and maximum values of burnup and fluence reached with AVR spherical fuel elements that contain high quality TRISO particles.

Type	Reload #	Burnup [% FIMA]		Fluence $(\times 10^{25}$ neutrons/m^2, E>16 fJ)	
		Average	Maximum	Average	Maximum
HEU GO2	15	16.0	19	4.0	5.7
	20	7.5	11	0.8	1.8
LEU GLE3	19	8.5	10	2.2	3.0
LEU GLE4	21	11.0	14	1.7	2.6
	21-2	3.5	3.8	0.3	0.4

To determine fuel element temperatures in the AVR, a sophisticated, multi-element melt wire experiment was carried out. This experiment contained specially designed graphite matrix spheres which incorporated a set of 20 capsules, each containing a single melt-wire[15]. The melt-wires were fabricated of specific alloy composition that would melt if a precise temperature was exceeded. The monitoring spheres were added into the AVR core through standard fuel loading procedures. Upon discharge, they were X-rayed to assess the momentary maximum peak temperatures experienced during their passage through the AVR.

Based on the melt-wire measurement results, the underlying temperature distribution was extracted. Two normal distributions with peak temperatures of $1100 \pm 66°C$ and $1220 \pm 100°C$ were found that defined the variation in momentary maximum fuel element surface temperatures in the AVR between the inner core and outer core locations. Peak central fuel element temperatures were calculated to be $37.3°C$ and $75.6°C$ higher for the inner and outer locations, respectively. The resulting distributions of peak temperatures experienced by AVR fuel elements are shown in Figure 11. It should be noted that these very high temperatures were experienced at the very top of the core only for a short period on the elements passed through the core of the AVR.

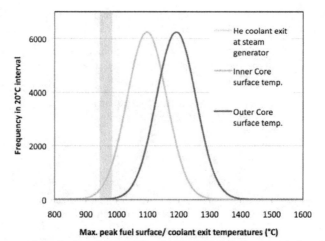

Figure 11: Two normal distributions (1100±66°C and 1220±100°C) define the variation of short-term maximum [at the pebble bed surface] fuel element surface temperatures in AVR between inner core and outer core. Fuel sphere center temperatures are 37/ 76°C higher.

PREDICTION OF IRRADIATION-INDUCED FAILED PARTICLES IN AVR TESTING

Analysis of in-reactor monitored R/B measurements to determine particle failure is not an applicable technique to detect particle failure for individual fuel elements in the AVR because only "total" core gas release rates are available during AVR operation. In addition, these measurements were dominated by dirty old fuel types.

To assess the EOL performance of AVR fuel elements, a methodology was developed that gives indications of AVR in-reactor particle failure based on analysis of the long-lived [85]Kr fission gas release measurements made during the gradual heatup in the early phase of accident condition testing[16,17,18]. This heatup process begins at room temperature, progresses over a series of heating-ramps to specific temperatures (300°C, 1050°C, and 1250°C) and hold periods until the desired simulation temperature is reached. Two of these hold points, 1050°C, and 1250°C, are designed to equilibrate the irradiated fuel particles in the fuel element at or near their prior irradiation temperature[19]. This allows the fuel to develop a stable internal environment before being heating to an elevated temperature, not previously experienced by the fuel particles. The 1050°C hold point was considered the mean working temperature for fuel specimens from accelerated MTR irradiation tests, and the 1250°C hold point was considered the typical working temperature for AVR fuel elements.

For the entire accident simulation test, the test furnace is purged with He sweep gas and continuously monitored for [85]Kr (10.76 yr half-life) release. Detection of any significant [85m]Kr activity in the sweep gas represents release from the fuel element and is an indicator of the presence of failed or defective fuel particles.

HEU (Th,U)O$_2$ TRISO Fuel In-AVR Performance

Based on the [85]Kr fractional release data from nine accident simulation tests, the EOL performance of the AVR GO2 fuel elements at their time of discharge from the AVR are excellent. Most of the fractional release data are <10^{-6} in the temperature range of 1250°C to well beyond 1800°C. The [85]Kr release fraction of a single HEU (Th,U)O$_2$ TRISO particle in an AVR GO2 element at these temperatures is $\sim 10^{-4}$. Collectively, the nine AVR GO2 fuel elements represent a population of 94,320 HEU (Th,U)O$_2$ TRISO coated particles with no observed in-reactor failure at discharge, Table 11. In similar accident condition testing performed on fuel elements with HEU (Th,U)O$_2$ TRISO particles, but irradiated in MTR irradiation experiments, the fractional release data compare well with that for the AVR GO2 elements.

Table 11: Noble gas ^{85}Kr release fractions measured during the heatup phase in accident simulation tests on AVR irradiated HEU (Th,U)O$_2$ fuel elements and irradiated spheres from MTR tests [16,17,18,19].

Fuel Element* [AVR Sample No./ Specimen No.]	Burnup (% FIMA)	Measurement Temperature (°C) of ^{85}Kr Release	^{85}Kr Release Fraction	Peak Temperature (°C) of Accident Simulation Test
KÜFA Isothermal Accident Simulation Tests				
AVR 70/26	8.2	1610**	$\leq 1.0 \times 10^{-6}$	1610
R2-K13/1	10.3	1250	$\leq 3.4 \times 10^{-7}$	1600
Graphite Furnace Tests				
AVR 70/15	7.1	1250	$\leq 7.0 \times 10^{-7}$	1500
AVR 70/7	7.3	1500**	$\leq 6.3 \times 10^{-7}$	1500
AVR 69/13	8.6	1800**	$\leq 5.4 \times 10^{-7}$	1800
AVR 74/24	11.2	1250	$\leq 5.4 \times 10^{-7}$	2100
AVR 74/20	11.9	1250	$\leq 1.6 \times 10^{-7}$	1900
FRJ2-K11/3	10.0	1600**	$\leq 5.1 \times 10^{-6}$	1600
Ramp Accident Simulation Tests in Graphite Furnace				
AVR 69/28	6.8	1530**	$\leq 6.8 \times 10^{-7}$	2150
AVR 70/18	7.1	2130**	$\leq 6.5 \times 10^{-6}$	2400
AVR 74/17	10.3	1250	$\leq 1.4 \times 10^{-7}$	2500

* The AVR Sample Number. represents the sequential sample of elements withdrawn for the AVR core for surveillance purposes; the Specimen Number is the order in which this element was withdrawn.
** No detectable release at 1250°C.

LEU UO$_2$ TRISO Fuel In-AVR Performance

A significant number of GLE3 fuel elements withdrawn from the AVR were subjected to accident simulation testing. From this testing, the results from 24 AVR elements were analyzed to determine the irradiation performance for the LEU UO$_2$ TRISO Fuel at the time of discharge from the AVR. Five of the tests were excluded because of the lack of detailed ^{85}Kr release data at testing temperatures near those expected in the AVR. The ^{85}Kr fractional release for the LEU UO$_2$ TRISO Fuel elements are presented in Table 12. The ^{85}Kr fractional release data obtained from six MTR irradiated elements with the same type of fuel particles are also presented. The MTR data are shown for comparison and are in good agreement with the data from the AVR irradiated elements.

Table 12: Noble gas ^{85}Kr release fraction measurements made during accident simulation tests on AVR type GLE 3 fuel elements. Similar results obtained on accelerated MTR tests HFR-K3 and FRJ2-K13 shown for comparison.

Fuel Element [AVR Sample No./ Specimen No.]	Burnup (% FIMA)	Measurement Temperature of ^{85}Kr Release	^{85}Kr Release Fraction	Peak Temperature (°C) of Accident Simulation Test
Isothermal Accident Simulation Tests				
AVR 70/33	1.6	1250	3.8×10^{-6}	1800
AVR 73/21 (ITU*)	2.5	NI**	NDR† <1600	1800
AVR 71/22	3.5	1250	4.5×10^{-8}	1600
AVR 74/18 (ITU*)	4.8	1600	5.9×10^{-6}	1600
AVR 74/10	5.5	1250	$< 8.0 \times 10^{-7}$	1800
AVR 74/11	6.2	1250	1.3×10^{-7}	1700
AVR 76/18	7.1	1250	1.6×10^{-8}	1800
AVR 88/41	7.6	1250	1.3×10^{-8}	1800
AVR 88/33	8.5	1250	$< 4.2 \times 10^{-8}$	1600/1800
AVR 82/20	8.6	1250	5.5×10^{-8}	1600
AVR 88/15	8.7	1250	6.3×10^{-8}	1600/1800
AVR 82/9	8.9	1250	1.3×10^{-8}	1600
HTR-Module Depressurized Loss-of-Coolant Event Profile				
AVR 91/31	9.0	1500	1.0×10^{-8}	1700
AVR 89/13	9.1	1490	6.7×10^{-8}	1620
AVR 85/18	9.2	1495	2.6×10^{-8}	1620
AVR 90/5	9.2	1495	5.3×10^{-8}	1620
AVR 90/2	9.3	1495	9.2×10^{-8}	1620
AVR 90/20	9.8	1500	5.7×10^{-8}	1620
Ramp Accident Simulation Tests				
AVR 71/7	1.8	1250	3.1×10^{-7}	2000
AVR 70/19	2.2	1200	1.9×10^{-6}	2400
AVR 74/8	2.9	1250	1.4×10^{-7}	2500
AVR 73/12	3.1	1250	$< 1.4 \times 10^{-7}$	1900
AVR 74/6	5.6	1250	1.3×10^{-7}	2100
AVR 76/28	6.9	1250/NI**	NDR† <1750	2100
AVR 76/19	7.3	1250	2.3×10^{-7}	1900
AVR 76/27	7.4	1250	4.6×10^{-7}	2100
AVR 80/16	7.8	1250/NI**	NDR† <1900	2000
AVR 80/14	8.4	1250/NI**	NDR† <1900	2500
AVR 80/22	9.1	1250/NI**	NDR† <1600	1900
Accident Simulation Tests on MTR and HTR Modul Proof Test Elements				
HFR-K3/1	7.7	1250	$< 5.6 \times 10^{-8}$	1600
HFR-K3/3	10.2	1250	1.5×10^{-7}	1800
FRJ2-K13/2	8.1	1250	5.3×10^{-7}	1600
FRJ2-K13/4	7.8	1250	4.5×10^{-8}	1600/1800
HFR-K6/2 (ITU*)	9.7	1050	1.0×10^{-8}	1600
HFR-K6/3 (ITU*)	9.8	1050	3.2×10^{-6}	1600

* Institute for Transuranium Elements (ITU), Joint Research Centre, Karlsruhe, Germany.
** NI = Not Included in EOL AVR Evaluation. †NDR = No Detectable Release

Based on the measured [85]Kr fractional release, the performance of the Type GLE 3 fuel elements at the time of discharge from the AVR is excellent. The fractional release data for all the GLE elements are <6 x10^{-6} in the temperature range of 1250°C to well beyond 1400°C. These release data are significantly less (>10X to >1000X) than the total [85]Kr of ~6 x10^{-5} generated in a single LEU UO$_2$ TRISO particle at these temperatures. Collectively, these 24 AVR GLE3 fuel elements represent a population of ~393,600 LEU UO$_2$ TRISO coated particles with no observed in-reactor failure at discharge.

A comparison of the [85]Kr fractional release measurements monitoring during the equilibration phase prior to accident simulation testing for AVR GO2 elements with HEU (Th,U)O$_2$ TRISO fuel and AVR GLE3 elements with LEU UO$_2$ TRISO fuels and is presented in Figure 12. This comparison shows there is no systematic difference in irradiation performance for fuel elements containing either the HEU (Th,U)O$_2$ TRISO or the LEU UO$_2$ TRISO fuel particles at the time of their discharge from the AVR.

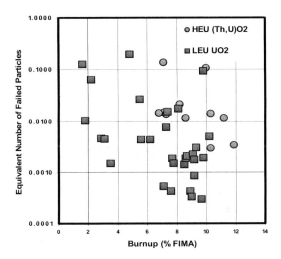

Figure 12: Fractional particle inventory from noble gas [85]Kr release monitoring during irradiation equilibration phase prior to accident simulation testing of AVR type GO2 fuel elements with HEU (Th,U)O$_2$ TRISO fuel and AVR GLE3 LEU UO$_2$ TRISO fuels.

IN-REACTOR PERFORMANCE COMPARISON BETWEEN HEU/TH TRISO AND LEU TRISO

The successful development and the irradiation performance results for the LEU UO$_2$ fuel system within the German program has been well-documented[20,21]. The LEU UO$_2$ fuel system was licensed in Germany in the 1990's for the HTR Modul concept[11] and it remains today a viable fuel concept for all HTR applications.

A comparison between the in-reactor performance of the HEU (Th,U)O$_2$ TRISO fuel system and the LEU UO$_2$ TRISO fuel system is provided in Table 13. Since both fuel systems were subjected to similar development efforts, this comparison is straightforward. Both HEU (Th,U)O$_2$ and the LEU UO$_2$ fuel systems were tested in qualification tests carried out in European MTRs and both were the subject of two large fabrication and irradiation campaigns in the AVR. Comparing the operating conditions during irradiation shows a similar burnup and accumulated fast fluence range. The major difference was in peak operating temperatures in that the MTR tests ranged from 1200°C to 1600°C for the HEU (Th,U)O$_2$ fuels, compared to the 1100°C to 1200°C for the LEU UO$_2$ fuels. This operating temperature difference may explain why the (Th,U)O$_2$ fuels have a higher in-reactor failure level of 6.5 x10^{-5} as compared to 2.1 x10^{-5} for the LEU UO$_2$ fuels. However, the performance results for both fuel systems are extremely good and represent world-wide performance standards.

Table 13: Comparison of in-reactor fuel performance between the HEU (Th,U)O$_2$ TRISO and the LEU UO$_2$ TRISO fuel systems.

Standard HTR Spherical Elements and Fuel Bodies with Oxide TRISO particles		No. fuel bodies	No. coated particles N	No. in-reactor failed particles n	Expected failure fraction =n/N	Upper 95% confidence limit
LEU UO$_2$ TRISO						
Irradiation Testing	MTRs Standard	19	276,680	0		
	Nonstandard	45	80,572	9		
	Total	64	357,252	9		
	AVR	24	393,600	0		
	Total	88	750,852	9	1.2x10^{-5}	2.1x10^{-5}
HEU (Th,U)O$_2$ TRISO						
Irradiation Testing	MTRs Standard	6	82,720	3		
	Nonstandard	24	43,950	5		
	Total	30	126.670	8		
	AVR	9	94,320	0		
	Total	42	220,990	8	3.6 x10^{-5}	6.5 x10^{-5}

ACCIDENT SIMULATION TESTING
Fuel temperatures in a depressurized-core-heatup accident scenario are sufficiently high to bring about a nuclear shutdown because of the large the negative temperature coefficient in the core. However, afterheat production resulting from fission product decay, in combination with a loss of forced circulation may lead to an unrestricted core heatup. For large HTR plants, such as the PNP and HHT design concepts, fuel temperatures can go beyond 2500°C. In contrast, the smaller modular HTR concepts, with their tall, small diameter, core automatically limits the maximum temperature to ~1620°C. The coolant in both the large PNP/HHT plants and in the modular HTRs is still helium.

ACCIDENT SIMULATION TESTING FACILITIES
These high temperature accident scenarios were simulated in the FZ-Jülich hot cell test furnaces beginning in the 1970s through the 1990s. Accident simulation was accomplished by externally heating irradiated spherical fuel elements under a purged helium environment in specialized test facilities. The He purge circuit incorporated the capability for continuous on-line measurement of ^{85}Kr release from the heated fuel in external cold traps.

Prior to 1984, a graphite high-temperature furnace (designated "A Furnace"), Figure 13, was the primary furnace employed and provided the capability of heatup ramps to a peak temperature of 2500°C. Continuous fission gas release monitoring was possible; however, the release of key solid fission product inventories could only be estimated by measurements before and after the accident simulation test.

Beginning in the mid-1980s, irradiated fuel elements containing high-quality HEU (Th,U)O₂ and LEU UO₂ TRISO fuel particles were becoming available for experimental post-irradiation evaluation. With the emphasis on passive safety for small, modular HTR concepts it became necessary to design/construct a new high-temperature heating facility to demonstrate these safety characteristics. This facility would incorporate state-of-the-art fission product detection capabilities to quantify gaseous and solid fission product behavior during accident condition testing.

Figure 14 is a schematic of the tantalum furnace design with a temperature limit of 1800°C and it is equipped with a built-in cold-finger apparatus (KÜFA=KühlFingerAnlage). Inclusion of the cold-finger assembly adds the capability of a semi-continuous measurement of solid fission product release in addition to the continuous gas release without having to interrupt the heating test. Figure 15 is a schematic of the operation of the fission product detection capabilities of the FÜFA Facility and the handling of the deposition plates.

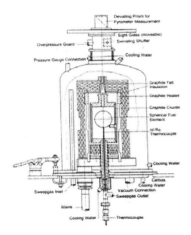 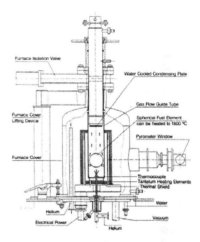

Figure 13: "A Furnace" graphite heating furnace, designated for performing accident condition tests at temperatures to 2500°C. Solid fission product release is determined by inventory difference before and after heating.

Figure 14: The KÜFA: a tantalum heating furnace with a water cooled cold-finger for accident simulations testing to 1800°C. Developed at the FZ-Jülich and is operational at the EU Institute of Transuranium in Karlsruhe, Germany.

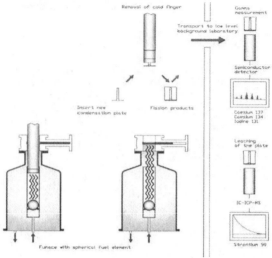

Figure 15: KÜFA operating diagram - Solid fission products released from a heating irradiated fuel element are deposited *in situ* on a cooled steel condensation plate. The plate can be exchanged regularly without interrupting the test. Solid fission product inventories on the plates are quantitatively assayed using gamma-spectrometry and mass-spectrometry [beta emitters] outside the Hot Cell[22,23,24].

ACCIDENT TESTING METHODOLOGY

The methodology used during accident testing simulation was developed at the Forschungszentrum Jülich. Historically two types of heating profiles were employed – one for ramp heating tests and one for isothermal heating tests. In Figure 16 the initial heatup phase to 1250°C , common to both types of test types, along with ramp testing profile out to 2500°C and the isothermal testing profiles at 1600°C to 1800°C are displayed. The hold points at 1050°C and 1250°C are for equilibration to readjust the fuel to its prior irradiation operating temperature conditions.

Any fission gas (i.e. ^{85}Kr) released during the accident simulation test is trapped in cold traps located in the helium purge gas circuit and quantitatively assayed using gamma-spectrometry. Solid fission product release (i.e. key fission products ^{90}Sr, ^{134}Cs, ^{137}Cs, and ^{110m}Ag) is measured in two different ways dependent upon the heating test facility employed. In accident simulation tests conducted in the KÜFA facility (Figure 13), solid fission products plate out on a cooled condensation plate which can be replaced at regular intervals without interrupting the heating process. Once a plate is removed/replaced on the cold finger of the KÜFA furnace, it is unloaded from the hot-cell and the solid fission product inventories quantitatively assayed using gamma-spectrometry and ICP mass-spectrometry [beta emitters]. In tests performed in the "A Furnace" facility, the losses of key solid fission products are estimated based on inventory measurements made by gamma spectrometry on the intact spherical fuel element before and after the heating procedure.

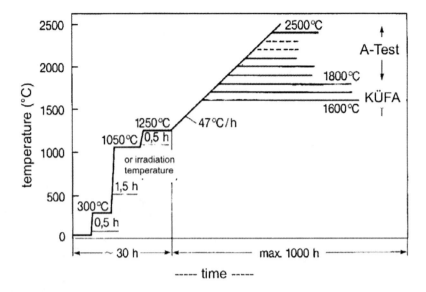

Figure 16: Typical temperature/ time profile in the ramp up to isothermal accident condition heating tests at 1600 and/or 1800°C. The 1250°C equilibration period can be used to determine the particle failure level at the end of the prior irradiation.

OBSERVATIONS FROM THE RAMP TESTS, 1900°C TO 2500°C

Accident simulation tests with ramp temperature profiles have been performed on three irradiated AVR GO2 fuel elements (Table 11). The ^{85}Kr fission gas release fractions were in the ~6 x10^{-7} to <3 x10^{-6} range, far below the fractional release representing the inventory of a single HEU (Th,U)O$_2$ particle. Fission gas monitoring up to 2050°C indicating no HEU (Th,U)O$_2$ TRISO particle failure. Figure 17 is a comparison of ^{85}Kr fission gas release data for two of the AVR GO2 elements (AVR 70/18 and AVR 74/17) with similar data from accident simulation ramp tests on AVR GLE3 fuel elements (AVR 70/19 and AVR 74/8). The latter two elements contained LEU UO$_2$ TRISO particles. The ^{85}Kr fractional release data, as a function of heating temperature, are very comparable for all four AVR irradiated elements. The GO2 elements with HEU (Th,U)O$_2$ fuels had a significantly higher burnup - 7.1% and 10.3% FIMA - compared to 2.2% and 2.9% FIMA for the LEU UO$_2$ fueled elements. No particle failure was indicated in any of the four elements tested up to temperatures >2050°C.

As heating temperatures exceeds ~2100°C in accident simulation ramp tests, the ^{85}Kr fractional release curves make a dramatic shift upward and continue to rise as the temperature increases. Above 2100°C the release curve quickly reaches a level indicative of catastrophic particle failure. As shown in Figure 17 and Figure 1, this occurs in both HEU (Th,U)O$_2$ TRISO fuel and LEU UO$_2$ TRISO fuel at nearly the same temperature. This dramatic change in performance is due to a serious deterioration of

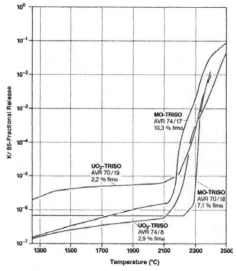

Figure 17: Comparison[19] of the [85]Kr fractional release from irradiated AVR GO2 [HEU (Th,U)O$_2$ TRISO] and AVR GLE3 [LEU UO$_2$ TRISO] fuel elements as a function of temperature during accident simulation ramp tests.

the TRISO coatings of fuel particles caused by the onset of SiC thermal decomposition. At temperatures >1800°C, the SiC layer will readily decompose[25,26,27] into its constituent elements

$$SiC(s) \rightarrow Si(g) + C(s) \qquad \text{Equation 2}$$

where the Si (g) vaporizes away leaving a porous carbon material behind.

At HTR normal operating temperatures, the SiC decomposition rate is negligibly small and is not a contributor to particle failure. Even at anticipated accident conditions of 1620°C for several hundred hours, SiC decomposition is insignificant. Visible deterioration of the SiC layer in the TRISO coating is recognized at temperatures >1800°C, caused by the onset of SiC thermal decomposition.

OBSERVATIONS FROM THE ISOTHERMAL TESTS, 1600°C TO 1800°C
The initial irradiated fuel elements available for evaluation in the KÜFA facility (1984) were contained HEU (Th,U)O$_2$ TRISO fuel particles. Two of the early 1600°C isothermal heating tests were performed on the GO2 fuel element AVR 76/20 and the fuel element R2-K13/1.
The fractional release results[17] for the key fission products [85]Kr, ^{90}Sr, ^{137}Cs, and [110m]Ag monitored in the R2-K13/1 and AVR 70/26 accident simulation tests are shown in Figure 18 as a function of heating time at 1600°C. The [85]Kr fractional release for AVR 70/26 and R2-K13/1 remained <10^{-5} well beyond 200 hours at 1600°C and are significantly lower than for other AVR GO2 elements subjected to accident simulation testing, Figure 19. Elements AVR 70/7, AVR 69/13, AVR 70/15 and AVR 74/24 were subjected to testing in the "A-Furnace" facility, and elements AVR 70/26 and R2-

K13/1 in the KÜFA facility. The higher release fractions may be a result of uncertainty in furnace temperatures in the "A-Furnace" facility

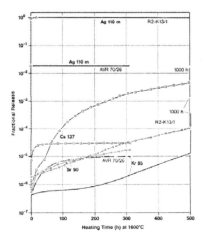

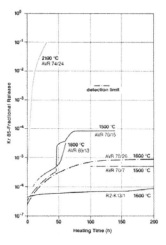

Figure 18: Fission product release fractions measured during 1000 hour (R2-K13/1) and 312 hour (AVR 70/26) heatup tests[17] at 1600°C from fuel elements with HEU (Th,U)O$_2$ TRISO particles.

Figure 19: Noble gas ^{85}Kr fractional release[17] as a function of isothermal heating time for irradiated AVR GO2 elements and MTR irradiated element R2-K13/1.

The ^{137}Cs fractional release data monitored for AVR 76/20 and R2-K13/1 as a function of heating time at temperature are shown in Figure 20 along with the profiles from other AVR GO2 element tested in the "A-Furnace" facility. The AVR 76/20 and R2-K13/1 elements were tested in the KÜFA facility and their ^{137}Cs release profiles are quite different from the very beginning of the isothermal heating phase. The ^{137}Cs release profile from element AVR 76/20 rises initially at the beginning of the 1600°C heating phase and then flattens out quickly after ~30 hours of heating. For the element R2-K13/1, the ^{137}Cs fractional release profile starts out at a ~10X lower value than for AVR 76/20 and begins to rise immediately with the 1600°C heating phase. After ~60 hours of heating at 1600°C, the R2-K13/1 ^{137}Cs fractional release exceeds that of the AVR 76/20 element. This increase continues throughout the heating phase. At ~95 hours the R2-K13/1 ^{137}Cs fractional release is equivalent to the inventory of two HEU (Th,U)O$_2$ TRISO particles, and after 230 hours it is equivalent to ~20 particles.

The ^{137}Cs fractional release rate data plotted in Figure 21 illustrates the different phenomena that are occurring in the AVR 76/20 and R2-K13/1 fuel elements while undergoing the same type of accident simulation testing. The shapes of the two release rate curves shown here are indicative of the different behavior of the HEU (Th,U)O$_2$ TRISO fuel particles in the two irradiated fuel elements. The lower ^{137}Cs curve for AVR 76/20 is representative of a depleting source for cesium, and conversely, the upper ^{137}Cs release curve for R2-K13/1 is characteristic of an increasing cesium source.

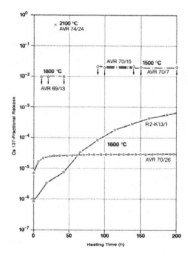 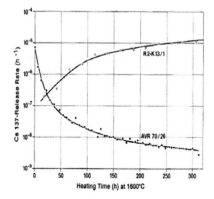

Figure 20: Fractional release[17] of ^{137}Cs from fuel elements AVR 76/20 and R2-K13/S with HEU (Th,U)O$_2$ TRISO fuel particles.

Figure 21: Fractional release rate for ^{137}Cs measured during accident simulation testing[17] of irradiated fuel elements AVR 76/20 and R2-K13/1

The depleting cesium source in the AVR 76/20 fuel element is located in the fuel matrix. This source term is a result of as-fabricated heavy metal contamination and additional cesium contamination picked up from years of service in the AVR and they are both being driven off by the high temperature heating process. As the isothermal heating proceeds, the A3-27 fuel matrix continues to loose cesium through this purification process. This behavior illustrates that the HEU (Th,U)O$_2$ TRISO fuel particles in AVR 76/20 remain intact, retaining their ^{137}Cs inventory, not allowing any diffusive release through the TRISO coatings during heating. The exact opposite is happening in the R2-K13/Capsule 1 fuel element. The increasing source term is a result of diffusive ^{137}Cs release through the SiC and PyC layers in the TRISO coatings of HEU (Th,U)O$_2$ fuel particles even though the particles appear to be intact as there is no indication of increased ^{85}Kr release or any bursts of activity.

Two different HEU (Th,U)O$_2$ TRISO particle batches were tested in these two elements. The most significant characterization difference between the AVR 76/20 element is that it contains a particle batch with a Th/^{235}U ratio of 5.00, while the R2-K13/1 element has a Th/^{235}U ratio of 10.02. The other important fact is that their irradiation conditions were different in that the operating temperatures in AVR 76/20 were probably on average lower than in the R2-K13 experiment because of the cycling effect in the AVR. Further, the accumulated fluence in AVR 76/20 was considerably lower at 2.0 x10^{25} n/m^2 compared to 8.3 x10^{25} n/m^2, E>16fJ in R2-K13/1.

LEU UO$_2$ LTI TRISO ACCIDENT CONDITIONS TESTING

In the period 1985 to 1994, a number accident simulation tests were executed on irradiated spherical elements containing LEU UO$_2$ TRISO fuel particles, including: 14 isothermal tests in the range of 1600° to 1800°C in the KÜFA facility[17,18,28]; six HTR Modul short-term heatup tests to 1600°C and 1700°C in the KÜFA facility[17,18,29]; and 11 ramp tests up to maximum temperatures of 1900°C to 2500°C . The results from 20 of these tests conducted in the KÜFA facility, four MTR irradiated fuel elements (from HFR-K3 and FRJ2-K13) and 16 GLE3 fuel elements irradiated in the AVR, are listed in Table 14. This table identifies the elements tested, irradiation history, accident temperature simulation/duration, and the total fission product fractional release results during the accident simulation testing. The 11 accident simulation ramp tests were listed previously in Table 12.

.

OBSERVATIONS FROM THE RAMP TESTS, 1900 °C TO 2500°C

The ^{85}Kr fractional release results monitored from ten of the eleven accident conditions ramp tests with AVR[18] GLE3 fuel elements are displayed in Figure 22 over the temperature range from <1300°C to 2500°C. The release fractions, with the exception of AVR GLE3 element 80/22, all remain <6x10^{-6} to temperatures >1700°C. For element 80/22, the fractional release did not increase until somewhere below ~1600°C, at which point a dramatic increase occurred due to failure of a number of LEU UO$_2$ TRISO particles. Based on the ^{85}Kr fractional release of 4.0x10^{-4} recorded at 1600°C, this translates into ~7 to 13 equivalent failed LEU UO$_2$ particles. With the exception of element 80/22, the fractional release results for the other nine elements shown in Figure 22 are far below the release fraction representative of a single LEU UO$_2$ TRISO particle of ~6x10^{-5}. All 11 accident simulation ramp tests were conducted in the "A-Furnace" facility at the FZ-Jülich, and as such are subject to the furnace operating temperature uncertainty associated with this facility.

Table 14: Results of 1600-1800°C accident simulation tests with irradiated fuel elements containing LEU UO$_2$ TRISO particle fuel in KÜFA tests performed at the Forschungszentrum Jülich.

Fuel Element	Burnup %FIMA	Fast Fluence 10^{25}m^{-2} E>16fJ	Heating test		Number of failed particles **		Fractional release				
			Temp (°C)	Time (h)	manuf.	heating	^{85}Kr	^{90}Sr	^{113m}Ag	^{134}Cs	^{137}Cs
AVR 71/22	3.5	0.9	1600	500	no	no	4.0E-7	5.3E-6	9.0E-4	6.9E-5	2.0E-5
HFR-K3/1	7.7	3.9	1600	500	no	no	1.8E-6	1.8E-7	2.7E-2	1.3E-4	1.1E-4
FRJ2-K13/2	8.0	0.1	1600	160	no	no	6.4E-7	3.3E-7	2.8E-3	1.0E-4	3.9E-5
AVR 82/20	8.6	2.4	1600	100	no	no	1.5E-7	3.8E-6	4.4E-3	1.2E-4	6.2E-5
AVR 82/9	8.9	2.5	1600	500	no	no	5.3E-7	8.3E-5	1.9E-2	5.9E-4	7.6E-4
AVR 89/13	9.1	2.6	1620 *	~10	no	no	2.0E-7	***	8.3E-4	1.3E-5	1.1E-5
			1620 *	~10		no	1.3E-9	***	1.5E-2	1.6E-6	1.4E-6
AVR 85/18	9.2	2.6	1620 *	~10	no	no	1.4E-7	***	6.5E-3	1.0E-5	1.3E-5
AVR 90/5	9.2	2.7	1620 *	~10	no	no	1.9E-7	***	1.1E-3	7.7E-6	9.0E-6
			1620 *	~10		no	6.6E-9	***	9.0E-4	3.5E-6	3.3E-6
AVR 90/2	9.3	2.7	1620 *	~10	1	2	1.0E-4	***	3.7E-2	5.0E-5	4.6E-5
AVR 90/20	9.8	2.9	1620 *	~10	2	3	2.4E-4	***	7.6E-2	5.6E-6	6.5E-6
AVR 91/31	9.0	2.6	1700 *	~10	2	18	1.2E-3	***	6.2E-1	3.7E-3	2.4E-3
AVR 74/11	6.2	1.6	1700	185	1	no	3.0E-5	8.3E-5	3.2E-2	8.4E-5	7.6E-5
FRJ2-K13/4	7.6	0.1	1600	138	no	no	3.0E-7	2.0E-8	4.5E-4	5.7E-6	2.5E-6
			1800	100		2	7.2E-5	1.4E-3	5.3E-1	9.7E-3	9.9E-3
AVR 88/33	8.5	2.3	1600	50	no	no	1.0E-7	8.4E-6	1.2E-3	1.1E-4	1.2E-4
			1800	20		~4	1.8E-4	2.3E-4	2.1E-1	4.4E-4	4.6E-4
AVR 88/15	8.7	2.4	1600	50		no	6.3E-8	***	9.1E-3	8.8E-6	1.2E-5
			1800	50	1	~6	2.9E-4	1.1E-2	8.1E-1	1.3E-2	1.4E-2
AVR 70/33	1.6	0.4	1800	175	no	28	1.7E-3	***	***	***	2.2E-2
AVR 74/10	5.5	1.4	1800	90	no	30	81.2E-3	***	***	8.5E-2	7.9E-2
AVR 76/18	7.1	1.9	1800	200	no	~3	1.2E-4	6.6E-2	6.2E-1	5.3E-2	4.5E-2
AVR 88/41	7.6	2.0	1800	24	no	no	2.4E-7	1.2E-4	7.7E-2	1.4E-4	1.5E-4
HFR-K3/3	10.2	6.0	1800	100	no	~12	6.5E-4	1.5E-3	6.7E-1	6.4E-2	5.9E-2

* simulating calculated core heatup curve   ** out of 16,400 particles   *** not measured

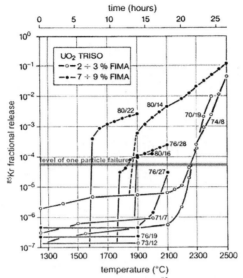

Figure 22: Release fraction for the noble gas fission product ^{85}Kr as a function of temperature measured during accident simulation ramp tests on AVR GLE3 fuel elements irradiated to various burnups in the AVR[18].

OBSERVATIONS FROM THE ISOTHERMAL TESTS, 1600°C TO 1800°C

Fourteen fuel elements, containing modern LEU UO_2 TRISO-coated fuel particles listed in Table 14, were subjected to isothermal accident simulation tests. The isothermal accident tests were distributed as follows:

- five 1600°C heatup tests (three AVR GLE3 and two MTR elements);
- three 1600°C heatup test followed by a 1800°C heatup test (two AVR GLE3 and one MTR elements;
- One 1700°C heatup test (one AVR GLE3 element); and
- Five 1800°C heatup tests (four AVR GLE3 and one MTR elements).

Figure 23 and Figure 24 present the fission product fractional release results for nine of the KÜFA isothermal heating tests performed in the FZ-Jülich. No single particle failure, nor any noticeable cesium or strontium releases, were observed during the first few hundred hours in any of the 1600°C heating tests. The measured time-dependent krypton and cesium fractional release profiles for those heating tests above 1600°C clearly demonstrates that particle failures and fission product release increase significantly as the accident temperatures rise. The 1700°C, 1800°C and 2100°C heatup tests explicitly demonstrate these points. With krypton and cesium as representative for a whole series of fission products, there is full retention at 1600°C for the accident specific first hundred hours or more (with ^{110m}Ag being an exception). In particular:

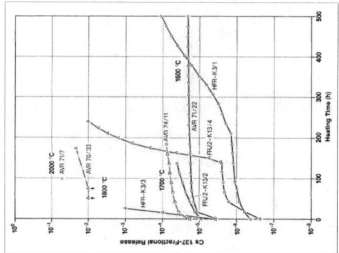

Figure 24: Cesium release profiles measured during isothermal accident simulation tests in the KÜFA with irradiated LEU UO$_2$ TRISO spherical fuel elements at 1600 to 2000°C[18].

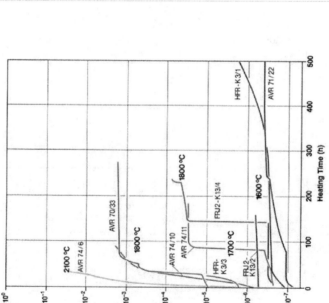

Figure 23: Krypton release[17] during isothermal accident simulation tests in the KÜFA with irradiated LEU UO$_2$ TRISO spherical fuel elements at 1600° to 2100°C. The ^{85}Kr fractional release curves indicates zero particle failures at 1600°C and one single particle failure at 1700°C.

- cesium is retained at 1600°C in the fuel particle and by the A3 matrix of the fuel element, with SiC providing for the primary retention barrier. This level of retention, however, has only been demonstrated with modern (after 1980) high quality TRISO coatings. At accident temperatures approaching 1800°C, there is no delay provided by the oxide fuel kernel and the graphite composite matrix. At these accident temperatures the SiC also becomes more permeable to most fission products due to the SiC decomposition phenomenon.

- the fission gas krypton is always released later than cesium, because of the additional retention capability provided by the dense, intact pyrocarbon layers. The exception is the sudden burst that accompanies rupture of the complete TRISO coating, commonly designated as a pressure vessel failure.

- strontium is more strongly retained in oxide fuel kernels and the graphite composite matrix than cesium. Therefore, strontium release generally occurs later than cesium, although its retention by SiC is not as good.

- even in the high quality SiC of modern TRISO coatings, silver is already released at irradiation temperatures above 1100°C. In some of the accident condition heating tests noted here, the ^{110m}Ag release fractions are approaching 100% in.

<u>OBSERVATIONS FROM THE HTR MODULE HEATUP SIMULATION TESTS, 1620°C & 1700°C</u>

Six HTR Modul accident simulation tests were executed and monitored in the KÜFA facility; five to 1620°C and one to 1700°C. These heating tests were identified in Table 12 and the fission product release results presented in Table 14. Figure 25 describes the ^{85}Kr fractional release as a function of heating time along with the prescribed heatup profiles at 1620°C and 1700°C. In two of the 1620°C tests, AVR GLE3 elements 90/2 and 90/20, two and three failed particles, respectively, were observed to have failed near the time when the temperature exceeded 1600°C. These particle failures are noted in Figure 25 by the stepwise increase in the ^{85}Kr fractional release at ~20 hours into the heating test. Likewise, for AVR GLE3 element 91/31, particle failure was first observed just as 1600°C was achieved as also noted in Figure 25 after ~12 hours of heating. Particles continued to fail in element 91/31 for a total of 18 until the maximum temperature of 1700°C was achieved. The three remaining AVR GLE3 elements, 89/13, 85/18, and 90/5 showed no indications of particle failure throughout the test as their ^{85}Kr fractional release values remained <2x10^{-7}. A total of five particle failures were observed in the five GLE3 elements (83,400 particles) subjected to the HTR Modul accident simulation heatup profile. and these are included in the 1600°C category of Table 16.

ACCIDENT CONDITION PERFORMANCE COMPARISON BETWEEN HEU (Th,U)O$_2$ TRISO and LEU UO$_2$ TRISO FUELS

The comparison between the accident condition performance of the HEU (Th,U)O$_2$ TRISO fuel system and the LEU UO$_2$ TRISO fuel system is provided in Table 13. All of these results summarized there are representative of accident simulation tests with peak temperatures ≤1620°C and were obtained on reference 60 mm-diameter spherical fuel elements. The disparity in the fuel populations tested reflects the emphasis placed on LEU UO$_2$ fuel development that began early in the early 1980s thru 1994. However, the performance between the two TRISO fuel particle types, ≤6.3x10^{-5} compared to ≤4.8x10^{-5}, is relatively consistent on a statistical basis even though ~5X more LEU UO$_2$ fuel was tested than HEU (Th,U)O$_2$ fuel. Compared to the design requirements for accident condition performance at ≤1600°C for both the HTR Modul and the PNP/HHT concepts at <2x10^{-4}, the performance of both TRISO fuel particle designs exceeds the requirement by >3X.

°General conclusions from evaluation of the accident simulation testing conducted on irradiated spherical fuel containing HEU (Th,U)O$_2$ and LEU UO$_2$ TRISO fuel particles are:

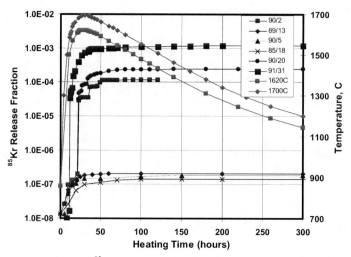

Figure 25: Fractional release of ^{85}Kr measured during the HTR Module accident simulation heatup tests conducted on seven GLE3 fuel elements irradiated in the AVR. Five tests were executed to a peak temperature of 1620°C and one to a peak temperature of ~1700°C.

Table 15: Comparison of performance under accident condition between the HEU (Th,U)O$_2$ TRISO and the LEU UO$_2$ TRISO fuel systems. Results presented are based upon accident simulation tests performed at the Forschungszentrum Jülich, Germany in the period 1980 to 1995.

Standard HTR Spherical Elements and Fuel Bodies with Oxide TRISO particles		No. fuel bodies	No. coated particles N	No. in-reactor failed particles n	Expected failure fraction =n/N	Upper 95% confidence limit
LEU UO$_2$ TRISO						
Accident Simulation Testing	Ramp Tests	11	183,480	~up to 13		
	Isothermal	8	133,440	0		
	HTR Modul	5	83,400	5		
	Total	24	400,320	18	4.28 x10^{-5}	6.35 x10^{-5}
HEU (Th,U)O$_2$ TRISO						
Accident Simulation Testing	Ramp Tests	3	31,440	0		
	Isothermal	2	30,530	0		
	Total	5	61,970	0	0	4.83 x10^{-5}

- AVR fuel elements (types GO2, GLE3 and GLE4) are soaked with cesium contamination on their surface. This contamination is a result of the prior irradiation of large number of releasing elements that remained in the AVR core from early fuel production campaigns. However, modern HTR fuel elements containing high-quality TRISO coated particles have demonstrated that they do not release cesium in long-term irradiations or during prolonged heating at temperatures $\leq 1600°C$. The relatively low average operating temperatures and low accumulated fast fluence in the AVR by comparison to spherical fuel elements irradiated in high-flux Material Test Reactors.

- Reference fuel elements irradiated in MTRs have significantly less contamination on their surface and release fission product at a much lower level as compared to the AVR irradiated elements. However, with continued heating, these MTR elements with their higher operating temperatures and higher accumulated fast fluences exhibit active diffusive release of solid fission products from intact TRISO particles during prolong heating at temperatures $\geq 1600°C$. Fission product retention is complete with the exception: silver for the first ~100 hours at 1600°C that are relevant for the HTR design basis accident scenario.

- Typically, the fractional release of krypton lags that of cesium because of the additional holdup provided by the intact PyC layers in TRISO coating design.

- The accident simulation ramp tests for both HEU $(Th,U)O_2$ TRISO and the LEU UO_2 TRISO fuel systems were executed in the "A-Furnace" facility at the Forschungszentrum Jülich. This facility has experienced on occasion loss of temperature control because of *in situ* decalibration of the temperature measurement system during test operation. Such loss-of-temperature control may contribute to over heating of one or more of the irradiated fuel elements under test.

- The performance results presented here are based on the assumption that the accumulated [85]Kr release fraction is an indicator of particle failure (~50% release at 1600°C and 100% release at 1800°C).

- The small number of HEU $(Th,U)O_2$ TRISO isothermal accident tests evaluated is due to unavailability of the KÜFA test facility at the FZ-Jülich until after 1984. Prior to this time all accident testing was carried out in the "A-Furnace" facility with a solid fission product detection limit of one to several percent of inventory. Prior results on AVR GO2 elements in 1400°C to 1800°C isothermal accident tests are unreliable due to uncertainty in test temperatures caused by repeated failure of the automated temperature control system.

More recently, KÜFA tests with irradiated LEU UO_2 TRISO spherical fuel elements have been performed at the EU commission Joint Research Center - Institute of Transuranium (ITU) in Karlsruhe, Germany. These tests are still under evaluation, but results clearly show no single particle failure in all of the Karlsruhe heating tests completed to date.

A careful analysis of released krypton, strontium, silver and cesium fission products allows a quantitative determination of the contribution from particle defects from manufacture and from particle failure induced during irradiation or during heating. With the statistical analysis of particle failure, it can be shown that the fuel failure during unrestricted core heatup does not significantly exceed the very low levels that exist from the defects in manufacture and the failures during irradiation.

PERFORMANCE ASSESSMENT FOR MODERN HTR TRISO FUEL SYSTEMS

With modern HTR fuels, present state-of-the-art requirements dictate:

- near complete retention of fission products at their source – the intact TRISO coated particles with no standard particle failure during normal operating conditions at temperatures <1250°C, and for accident conditions at temperatures ≤1600°C;
- very low levels of contamination in the outer PyC layer ($\leq 10^{-5}$) of the particle and in the fuel element graphitic matrix ($\sim 10^{-5}$); and
- low-levels of as-fabricated defective fuel particles ($\sim 10^{-5}$) with missing or defective coatings.

In this manner, the source term in an HTR is dominated by defective fuel particles produced during manufacture and only by their failure during irradiation or in accidents. Analysis of the many irradiation and accident conditions tests conducted between 1977 and 1994 have demonstrated excellent fuel behavior and their final performance assessment is limited only by sampling statistics. The performance statistics for both the HEU (Th,U)O$_2$ TRISO fuel system and the LEU UO$_2$ TRISO fuel system, illustrated in Table 16, are in perfect concert with the state-of-the-art requirements for present-day High Temperature Reactor concepts.

Table 16: Comparison of the performance statistics obtained on reference HTR 60 mm-diameter fuel elements containing HEU (Th,U)O$_2$ TRISO and LEU UO$_2$ TRISO fuel particles.

Standard HTR Fuel Elements	HEU (Th,U)O$_2$ TRISO Fuel			LEU UO$_2$ TRISO Fuel		
	No. coated particles	No. defect/failed particles	One-sided upper 95% confidence limit	No. coated particles	No. defect/failed particles	One-sided upper 95% confidence limit
Manufacture	487,480	5	$\leq 2.2 \times 10^{-5}$	2,202,200	86	$\leq 4.7 \times 10^{-5}$
Normal Operating Conditions	177,040	3	$\leq 4.4 \times 10^{-5}$	670,280	0	$\leq 4.5 \times 10^{-6}$
Accident Testing ≤1600°C	61,970	0	$\leq 4.8 \times 10^{-5}$	400,320	18	$\leq 6.3 \times 10^{-5}$

REFERENCES

1   Bäumer, R., Kalinowski, I., Röhler, E., Schöning, J., Wachholz, W. (1990), Construction and
     Operating Experience with the 300-MW THTR Nuclear Power Plant, Nucl. Eng. and Des. 121,
     pp.155-166.

2   R. Schulten, W. Bellermann, H. Braun, H.W. Schmidt, A. Setzwein, W. Stürmer, "Der
     Hochtemperaturreaktor von BBC/Krupp", Die Atomwirtschaft, 1959.

3   R. Schulten, "Pebble bed HTRs", Annals of Nuclear Energy 5 (1978), 357.

4   Fortescue, P., Bell. F.R., Duffield, R.B. (1965), Hexagonal Fuel Element, U.S. Patent
     Application No. 485,811, Filed September 8, 1965.

5   R.C. Dahlberg, R.F. Turner, W.V. Goedell, "Fort St. Vrain Core Design Characteristics", Nucl.
     Eng. Int. 14 (1969), 163.

6   M. Hrovat, H. Nickel, and K. Koizlik, "Über die Entwicklung eines Matrixmaterials zur
     Herstellung gepresster Brennelemente für Hochtemperatur-Reaktoren", Forschungszentrum
     Jülich Report Jül-969, June 1973.

7   M. Hrovat, H. Huschka, A.-W. Mehner, and W. Warzawa, "Spherical Fuel Elements for Small
     and Medium Sized HTRs", Nucl. Eng. Des. 109 (1988), 253.

8   K. Kugeler, M. Kugeler, H. Niessen and H. Hohn, "Design of a 3000 MW(th) High
     Temperature Reactor for Process Heat Applications", Nucl. Eng. Des. 34: 33-50 (1975).

9   HHT-NINT-Anlagenkonzept 1000 MW – Abschlussbericht, Hochtemperaturreaktor
     Heliumturbine Grosser Leistung, Band 1 (December 1977).

10   H. Reutler, G.H. Lohnert, "Advantages of Going Modular", Nucl. Eng. Des. 78 (1984), 129.

11   W. Heit, H. Nabielek, H. Ragoss and W. Schenk "Absicherung der brennelementspezifischen
     Quellterme für die radiologische Auslegung des HTR Modul", Jahrestagung Kerntechnik,
     Travemünde, June 1988 (Proc. p. 461ff).

12   K. Verfondern, "Fuel Performance and Fission Product Behavior in Gas Cooled Reactors",
     IAEA-TECDOC-978, Vienna, November 1997.

13   E. Sauer (ed.), "AVR-Experimental High Temperature Reactor: 21 Years of Successful
     Operation for a Future Technology", Association of German Engineers VDI, Düsseldorf,
     Germany, June 1990.

14   H. Nabielek, K. Verfondern, and M.J. Kania, "Fuel and Fission Products in the AVR Reactor",
     HTR2008, Washington D.C., 28 September - 1 October 2008.

15   H. Gottaut and K. Krüger, "Results of Experiments at the AVR Reactor", Nucl. Eng. Des. 121
     (1990), 143.

16   W. Schenk, "Störfallsimulation an bestrahlten Kugelbrennelementen bei Temperaturen von
     1400 bis 2500°C", Forschungszentrum Jülich Report Jül-1883, December 1983.

17   W. Schenk, D. Pitzer, and H. Nabielek, "Fission Product Release Profiles from Spherical HTR
     Fuel Elements at Accident Temperatures", Forschungszentrum Jülich Report Jül-2234,
     September 1988.

18   W. Schenk, and H. Nabielek, "Kugelbrennelement mit TRISO-Partikeln bei Störfalltemperaturen", Forschungszentrum Jülich Report Jül-Spez-487, January 1989.

19   W. Schenk, "Untersuchungen zum Verhalten von beschichteten Brennstoffteilchen und Kugelbrennelementen bei Störfalltemperaturen", Forschungszentrum Jülich Report Jül-1490, May 1978.

20   H. Nabielek, W. Kühnlein, W. Schenk, W. Heit, A. Christ, H. Ragoss, (1990) "Development of Advanced HTR Fuel Elements", Nucl. Eng. Des. 121 (1990), 199.

21   D. Petti, J. Maki, J. Hunn, P. Pappano, C. Barnes, J.J. Saurwein, S. Nagley, J. Kendall, and R. Hobbins, "NGNP Fuel Qualification White Paper", Idaho National Laboratory Report INL/EXT-10-18610, Rev. 0, 21 July 2010.

22   E.H. Toscano, W. Klopper, and J.H. Venter, "The Guidebook on Post-Irradiation Examination of HTR Fuel Elements", IAEA Document, November 2010.

23   D. Freis, "Störfallsimulationen und Nachbestrahlungsuntersuchungen an kugelförmigen Brennelementen für Hochtemperaturreaktoren", PhD Thesis University Aachen, 1 March 2010.

24   D. Freis, D. Bottomley, J. Ejton, W. de Weerd, H. Kostecka, and E. H. Toscano, "Postirradiation Testing of High Temperature Reactor Spherical Fuel Elements Under Accident Conditions", J. Eng. Gas Turbines Power 132, Issue 4 (2010), 042901 (6 pages).

25   K. Ikawa, F. Kobayashi, K. Iwamoto, "Failure of Coated Fuel Particles during Thermal Excursion above 2000°C", J. Nucl. Sci. Technol. 15 (1978), 774

26   R. Benz and A. Naoumidis, "Stability of Unirradiated Triso-Coated UC2-UO2 Particles with Different UC2 Contents at Elevated Temperatures", J. Nucl. Mat. 97 (1981), 15.

27   H. Nabielek, A. Naoumidis, D.T. Goodin, and K. Ikawa, "High Temperature Behavior of HTR Fuel Particles", Jahrestagung Kerntechnik, Berlin, 14-16 June 1983.

28   W. Schenk, R. Gontard, and H. Nabielek, "Performance of HTR Fuel Samples under High-Irradiations and Accident Simulation Conditions, with Emphasis on Test Capsules HFR-P4 and SL-P1", Forschungszentrum Jülich Report Jül-3373, April 1997.

29   W. Schenk, R. Gontard, and H. Nabielek, "Performance of HTR Fuel Samples under High-Irradiations and Accident Simulation Conditions, with Emphasis on Test Capsules HFR-P4 and SL-P1", Forschungszentrum Jülich Report Jül-3373, April 1997.

STUDY OF THE SILICON CARBIDE MATRIX ELABORATION BY FILM BOILING PROCESS

Aurélie Serre*, Joëlle Blein*, Yannick Pierre*, Patrick David*, Fabienne Audubert**, Sylvie Bonnamy***, Eric Bruneton*

*CEA, DAM, Le Ripault, 37260 Monts, France
**CEA, DEN, Cadarache, 13108 Saint-Paul-Lez-Durance, France
***CRMD, CNRS, Université d'Orléans, 45071 Orléans, France

ABSTRACT
    Film-boiling technique is a Ceramic Matrix Composites fabrication process, especially studied for the densification of carbon based composites. The densification starts from the porous preform, heated at high temperature in a liquid precursor which becomes gaseous in contact with hot surfaces. The cracking of the vapours results in a solid deposit constituting the matrix of the final composite. The main advantage of the film-boiling technique is its high densification rate: up to several hundreds of μm/h depending on the experimental set up (versus ~ 2μm/h in CVI). To get a better insight on this process, laboratory scale equipment with an internal resistive heating was developed. In this work, carbon preforms are densified by silicon carbide (SiC) matrices. Two SiC precursors are used: the methyltrichlorosilane (MTS) and the CVD4000. The samples characterizations show that the quality of the matrix (density, homogeneity, crystalline phase...) and the densification rate depend on several experimental settings (type of precursor, densification temperature...).

INTRODUCTION

    The Ceramic Matrix Composites (CMC) are tough ceramic materials comprising ceramic fibres, generally carbon or silicon carbide (SiC), embedded in a ceramic matrix. CMC are high performance materials although they are inherently constituted by brittle components. The carbon reinforced CMC are currently used in applications requiring light and refractory materials with a high wear and corrosion resistance, a good toughness and fatigue resistance, a low thermal dilatation and an excellent thermal shock resistance thanks to a good thermal conductivity. Major fields of applications are aerospatial and aeronautic parts such as disc brakes or thermal equipments[1,2].

    Various processes exist to manufacture thermostructural composite materials[3]. The Chemical Vapour Infiltration (CVI) is the most used process at the industrial scale: starting from a fibrous preform and a precursor, the matrix is generated by gas deposition[4]. Technical improvements resulted in various CVI processes which allow the manufacturing of C$_f$/C, C$_f$/SiC and SiC$_f$/SiC composites for several years[4-8]. The rapid densification or film-boiling method studied in this work was patented by Houdayer et al. in 1984[9]. This technique is based on the competition between chemical reactions and physical transport properties. It can be considered as an extreme case of thermal-gradient CVI (TG-CVI) for the densification of carbon and ceramic matrices. The specific equipment required for this process is called kalamazoo which is described in Figure 1[10].

    The main drawback is the lack of knowledge for the manufacturing of CMC by film boiling process: C$_f$/C composites have predominantly been studied[10-12]. A laboratory scale equipment or kalamazoo was used in previous works, it has an inductive heating (Figure 1) with a conductive coupling (graphite) allowing the densification of three dimensions preforms. But each densification needs several tens litres of precursor and several days of priming. An experimental device (called mini-kalamazoo) was therefore developed for the present parametric study which consists of testing different precursors to form SiC matrices by film-boiling method. The reactor being smaller, the quantity of precursor is clearly reduced (around 200 mL) as well as the vapour disposal during a densification. The time saving per experiment is also significant. But the heating is here resistive and implies a limitation in the porous preform choice which has to be a good electrical conductor. The carbon fibres T300 (Toray Carbon Fibres America, Inc.) were then used as fibrous preform. Six strands were braided by three in order to insure a sufficient specific surface, a good maintain of the fibres during the densification and the wedging of the thermocouple in the preform. The preform architecture is here not favourable to the infiltration but the study aims at evaluating different precursors in film-boiling process. The mini-kalamazoo allows dealing with various carbide depositions from different precursors, sometimes costly ones, reducing their consumption and the process time. The elaboration of three dimensional composites won't be performed in the present work but at a later time done on the bigger facility (the kalamazoo) in order to obtain infiltrated and dense composites.

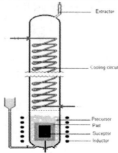

Figure 1. Schematic presentation of Kalamazoo using an inductive heating

EXPERIMENTAL

Process description

The film-boiling apparatus consists of two main parts: firstly the reaction chamber at atmospheric pressure which includes the reactor and the condenser, secondly the electrical power supply equipment. In the reactor, the fibrous structure to densify is immersed into a liquid precursor and heated by Joule effect at high temperature. Since the temperature of the part to densify is higher than the boiling temperature of the precursor, the precursor infiltrated into the fibrous structure is vaporized. In contact with the hot surfaces, the vapours are decomposed, resulting in the deposition of the solid matrix, and afterwards in the densification of the substrate. The densification occurs from the hottest zone (inside the fibrous preform) to the coldest one (towards the exterior of the preform). So, the plugging of the external pores is avoided, contrary to CVI or impregnation where the composite must be machined several times to densify inside. The thermal gradient inside the part (Figure 2) due to the highly endothermic vaporization phenomena, leads to convections which combined with the high chemical precursor concentration (thanks to its liquid state), form a fast densification front moving through the part[12]. The uncracked vapours are condensed within the cooling system, above the reactor, and the produced gases are released in the extractor.

The chemical kinetics is related to the gas phase and the heterogeneous surface reactions[13]. Then, the mass transport depends on the diffusion of the species as it is described by Fick's and Knudsen's laws[14] and the formation of the deposit results from the competition between the reaction and the diffusion mechanisms. The residence time of the gaseous species at a defined homogeneous deposition temperature also has an influence on the final microstructure of the material.

The film-boiling technique is not categorized as a CVI process because the precursor has to be a liquid, but it could be compared to the forced-flow-thermal gradient CVI. The reactive flow is initiated by the boiling liquid surrounding the fibrous preform. The deposition of the matrix is also a different mechanism than impregnation in liquid phase from organic agents. The rapid densification technique presents some advantages compared to classical processes. Indeed, the part can be manufactured in one step and the densification time is strongly reduced (50 to 600 times in comparison with CVI). Moreover, varying the types of precursor, different matrices can be obtained and generate various CMC.

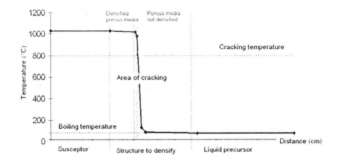

Figure 2. Example of temperature distribution and areas in a sample during the densification

Equipment and experimental set-up

The experimental device (Figure 3) hereinbefore presented allows testing expensive precursors in small quantities and the elaborated materials are at the "mini-composite" scale. The equipment has to be placed in inert atmosphere because of the moisture sensitive precursors. So, two gloves boxes under nitrogen atmosphere are connected by an airtight chamber. The pressure of nitrogen is kept slightly above 1 atm to prevent outside air from leaking into the glove box. The precursor is weighed in the first glove box and poured in a hermetically closed container. Then, it is transferred through the chamber into the other glove box which contains the densification system (reactor and cooling system).

The precursor is introduced into the reactor by a pumping system. The carbon fibres are locked between two copper electrodes and heated by Joule effect. The temperature of the carbon fibres is set with the intensity imposed by the power source and it is measured by a thermocouple wedged in the braided sample.

The first study leaded on the mini-kalamazoo concerned the elaboration of C_f/C mini-composites[15]. The results obtained were in accordance with the previous ones got on the kalamazoo. The mini-kalamazoo was thus approved to begin the study about SiC elaboration. Two SiC precursors have been tested: the methyltrichlorosilane or MTS (CH_3SiCl_3) and the hydridopolycarbosilane, also called CVD4000 ($[SiH_2-CH_2]_n$). MTS is a colourless liquid, sensitive to humidity with a boiling temperature equal to 66.4°C. This chemical compound produces chloric acid during its decomposition which requires specific safety precautions. MTS is often combined with methane and hydrogen to be used as SiC precursor in CVD or CVI[16-19]. On the contrary, CVD4000 is not common. Only one American patent[20] in the literature describes SiC films deposition using CVD4000 as precursor. CVD4000 is a SiC liquid precursor developed by Starfire Systems with a boiling temperature varying between 50 and 150°C. Since CVD4000 does not generate corrosive gases, it provides high purity films, free of impurities coming from the chloric acid (HCl) corrosion of metal parts in the equipment. The lack of chlorine in the CVD4000 molecule implies better equipment durability by inhibiting the formation of HCl in the system.

The main controlled experimental parameters are the densification time and the temperature which is regulated by the electrical intensity. For these experiments, its values range from 15 – 25 A, for the voltage they range from 15 – 30 V and 300 – 600 W for the power. The parameters chosen for this study are presented in Table 1.

Figure 3. Experimental device with a resistive heating: the mini-kalamazoo

In order to characterize the created matrices, several analysis techniques were used. The optical micrographs were performed with the Olympus GX71 on polished cross section samples. The coatings thicknesses were averaged on each sample thanks to ImageJ, a Java-based image processing program developed at the National Institutes of Health. The densification rates were then deducted by dividing the coating thickness by the deposition time.

Polished cross sections were also analysed by Scanning Electron Microscope (SEM) equipped with Energy Dispersive X-Ray Spectroscopy (EDS). The applied material was LEO 435 VPi. The SEM micrographs presented in this paper were performed with back scattered electrons. SEM/EDS technique gives information about the microstructure and the elementary composition of the samples.

The X-Ray Diffraction informs about the chemical composition and the crystal structure, it allowed us to know if SiC was present in the samples. The separation between the carbon fibres and the deposited matrices being impossible, they were milled together before being analysed with the Siemens Diffraktometer D5000. So the XRD pattern of the carbon fibres was presented and had to be considered in the interpretations of the XRD results.

The Electron Probe Micro-Analyzer (EPMA) has a better resolution than EDS. The composition and the distribution of the chemical compounds were deduced from this technique. The model used is equipped with four Wavelength Dispersive Spectrometers (WDS), it is the Cameca SX50.

A specific set up was built to perform gas analyses. Indeed, a stainless steel bottle is added on the evacuating way of the gas at the exit of the condenser to collect the gases produced during the densification. They are then analyzed by Fourier Transformed InfraRed spectroscopy (FTIR) with the Bruker IFS 55.

Table 1. Temperatures and times densification according to the employed precursor

Precursor	Densification temperature (°C)	Densification time (min)	Intensity increase
MTS	700	210	
	800	30	
		90	
		150	
	1000	30	
		90	
		150	
	1200	30	
		90	
		150	1 A/min
CVD4000	600	90	
	700	90	
	800	30	
		90	
		150	
	1000	30	
		90	
		150	
	1200	30	
		90	
		150	

RESULTS AND DISCUSSION

Thermodynamic considerations and gas phase analysis

A thermodynamic study was performed with Gemini software (Thermodata, Saint Martin d'Hères, France) and it is based on the minimization of the total Gibbs energy of the system under either constant pressure or volume conditions. It is important to remind that any kinetic factor is taken into account.

The condensed phases predicted by the thermodynamic during the thermolysis of MTS and CVD4000 are respectively shown on Figure 4 and Figure 5. The deposits elaborated from MTS could contain carbon and SiC. The C/SiC ratio would vary with the temperature and the carbon proportion would be maximum for a densification at 800°C. On the contrary, the solid phase created from CVD4000 could be only composed of SiC. Concerning the gaseous phases, the theoretical results are presented on Figure 6 and Figure 7. The thermolysis of CVD4000 theoretically produces hydrogen (H_2) of which the chemical amount would practically not evolve with the temperature. On the contrary, the gaseous decomposition of MTS changes according to the temperature. The majority species should be $SiCl_4$, CH_4, H_2 and above 600°C, HCl is formed and from this point, its chemical amount would increase almost linearly with the temperature. In order to compare the thermodynamic predictions with the experimental measurements, the gases produced through the densifications from MTS were analysed by FTIR. Several tests were done at 1000°C after 15 min, 90 min and 150 min of densification: the gaseous phase does not evolve with the densification time at a fixed temperature. The results are presented in Table 2. In further works, gas analyses will be performed at different temperatures (800°C and 1200°C). The comparison between Figure 6 and Table 2 shows that the thermodynamic predictions are not in accordance with the experimental measurements. It is not surprising as it is difficult to reach a thermodynamic equilibrium of the gaseous phases during the film-boiling process. On the other hand, the gaseous species experimentally identified are similar to the ones present during the SiC CVD process in the system {MTS/H$_2$}[21], plus the presence of two hydrocarbons (acetylene and ethylene). These two last species could be responsible of carbon formation in the SiC deposit. Referring to condensed phases announced by the thermodynamic and the experimental gases analysis, a carbon excess could be present in the deposits elaborated from MTS.

On-going works consist on studying in the same way the CVD4000 thermolysis decomposition to compare the results with the thermodynamic. But we can already conclude that from thermodynamical considerations, these two SiC precursors initiate different deposit compositions during a densification by film-boiling process.

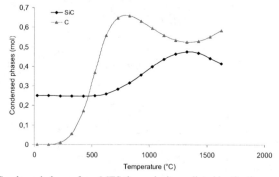

Figure 4. Condensed phases from MTS thermolysis predicted by the thermodynamic

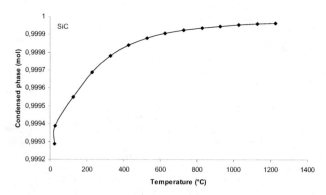

Figure 5. Condensed phase from CVD4000 thermolysis predicted by the thermodynamic

Table 2. Gaseous species identified by FTIR during
the thermolysis of MTS at 1000°C

HCl	Majority
CH$_4$	Majority
SiCl$_4$	Majority
MTS	Majority
C$_2$H$_2$	Minority
C$_2$H$_4$	Minority
SiCl$_2$	Minority
HSiCl$_3$	Minority

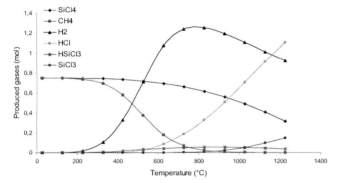

Figure 6. Gases from MTS thermolysis predicted by the thermodynamic

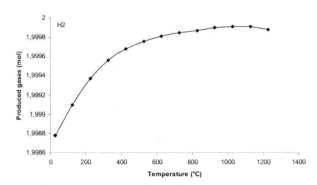

Figure 7. Gases from CVD4000 thermolysis predicted by the thermodynamic

Physico-chemical characterizations

The polished sections analyses with OM and SEM reveal that a deposit is formed from 800°C when MTS is used as precursor versus 600°C with CVD4000. In both cases, the substrate constituted by aligned carbon fibres is only infiltrated on a few micrometres on its periphery but as it is hereinbefore explained the conditions are not favourable to infiltration, the target being not the elaboration of dense composites on this device.

The general microstructure of the deposits is described on Figure 8 and Figure 9. The deposits made from MTS are dense without any cracks (Figure 8) contrary to the ones produced from CVD4000 (Figure 9). The cracks are strongly reduced by slowing down the temperature decrease at the end of the process (from 800°C/min to 50 °C/min). Moreover, it seems that in the same experimental conditions (same temperature and time densification), the matrix from CVD4000 is thicker than the one manufacture from MTS.

The SEM micrographs of the deposits elaborated by CVD4000 don't present any contrast and the EDS spectrums confirm that their composition is homogeneous: only silicon and carbon are present and their amounts are constant over the deposit. On the contrary, dark and clear zones appear on SEM micrographs for the samples elaborated from MTS. EDS analyses show a more intense carbon peak in the darkest zones than in the clearest, so they contain an excess of carbon. Nevertheless, the deposit around the fibres (~2 µm thick) is clear and the EDS signals of carbon and silicon are there constant (Figure 10). The thermodynamic predicts an excess of carbon maximum at 800°C but quantitative measurements are not possible from these experimental techniques, so we won't adjudicate this point. All the same, the results are in accordance with the condensed phase predicted by the thermodynamic and with the FTIR results demonstrating the presence of acetylene and ethylene. Indeed, these two identified hydrocarbons in the gaseous phase during the densification are favourable to the carbon formation. The SiC being formed on the substrate surface by a reaction between silicon and carbon respectively deposited from chlorosilane and hydrocarbon radicals, a gas phase enriched in hydrocarbons leads to a carbon excess in the deposit. The heterogeneous reactions on the substrate surface depend on the types and quantities of reactive intermediates species, so the identification of the gaseous phase during the densification helps to understand the deposit composition.

MTS is currently employed as SiC precursor in CVD and CVI processes and to avoid the excess of carbon hydrogen is added to MTS[22-24]. This alternative is not possible in the case of the film-boiling technique for safety reasons.

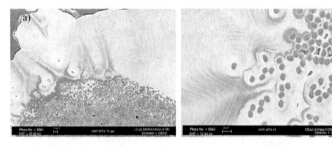

Figure 8. SEM micrographs of samples densified from MTS at 1000°C for a) 90 min and b) 30 min

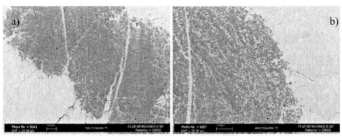

Figure 9. SEM micrographs of samples densified from CVD4000 at a) 600°C for 75 min and b) 700°C for 150 min

During the densification, the thermal conductivity of the whole sample varies because of the deposit growth, and leads to a thermal gradient which can be responsible of a composition gradient. So, the variances of the SiC/C ratio can also be a consequence of this phenomenon. In this way, several experiments were done surrounding the carbon fibres braids with a carbon felt which plays the role of thermal barrier and gives rise to a temperature homogenization over the substrate. The part of the felt in contact with the carbon fibres began to be densified (Figure 11), the composition of the deposit was then more homogeneous and the ratio C/SiC clearly reduced.

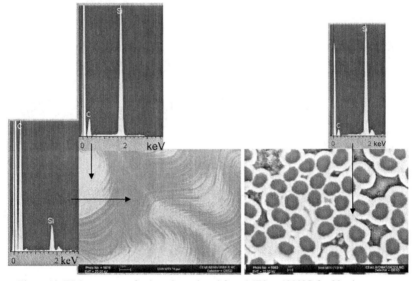

Figure 10. EDS spectrum of a deposit produced from MTS at 1200°C for 90 min

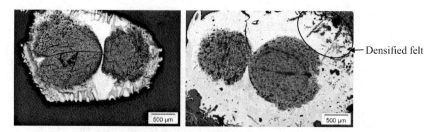

Figure 11. SEM micrographs of samples elaborated at 1200°C for 90 min a) without felt b) surrounded by a felt

The XRD pattern (Figure 12) confirms SiC formation in both cases (MTS and CVD4000). Two SiC polytypes are observed: α-SiC and β-SiC, identified by the JCPDS n°029-1131 and 029-1129 respectively. The ratio between these two SiC crystalline structures is impossible to deduce considering that their XRD patterns are overlaid. In this way, TEM analysis will be performed to distinguish α-SiC from β-SiC. The Figure 12 presents a diagram obtained from T300 fibres milled alone. Reminding that the whole sample is milled, the first large peak, more intense in the sample prepared with MTS than in the one prepared with CVD4000 corresponds to the carbon fibres. Then, the carbon excess cannot be underlined by XRD. The presence of silicon oxide (SiO_2) (JCPS n°046-1045) is observed even if all the experiments were performed under inert atmosphere. The deposits elaborated from MTS seem contain more SiO_2 than the ones from CVD4000. The oxygen generating the SiO_2 may come from oxidized carbon fibres, an oxidation during the process or a post-oxidation. In order to localize the SiO_2 in the samples, characterizations were performed by EPMA. The silicon, carbon and oxygen maps are presented on the Figure 13. The oxygen is mainly detected at the interface between the filaments and the deposit which confirms the hypothesis of carbon fibres oxidation. In addition, it can explain why SiO_2 was detected in the samples elaborated from MTS and not in the ones produced from CVD4000: in same conditions, a matrix from CVD4000 being thicker, the ratio carbon fibres/deposit is smaller than with MTS and the quantity of SiO_2 can be inferior to the detection limit of XRD (5 wt.%). In the following experiments, the carbon fibres underwent a thermal treatment at 400-500°C for 15 minutes in nitrogen atmosphere to remove the impurities and the peak assigned to SiO_2 was less intense.

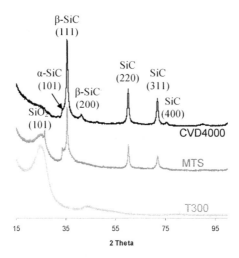

Figure 12. XRD patterns of milled samples (T300 carbon fibres alone, samples elaborated from MTS and from CVD4000 at 1200°C for 90 min)

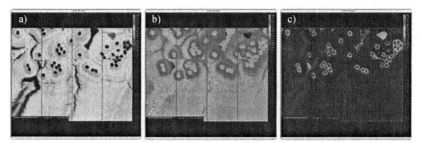

Figure 13. EPMA of a sample densified from MTS at 1200°C for 90 min
a) Si map, b) C map, c) O map

The precursor has an influence on the composition of the deposit. The process yields to SiC deposits with a carbon excess when MTS is used as precursor whereas pure SiC is got from CVD4000 but few cracks appeared in this last case. To conclude, the manufacturing of a SiC matrix is possible by film-boiling process.

Densification rate

Densification rates were calculated as mentioned before. The Figure 14 shows that the deposition rate depends on the temperature for both MTS and CVD4000. The densification rate increases exponentially with the fibrous preform temperature. For a defined temperature, the densification is faster using CVD4000 as precursor than MTS. Whatever the precursor and the temperature, the densification by film-boiling process is several times faster than CVI (50 to 600 times faster).

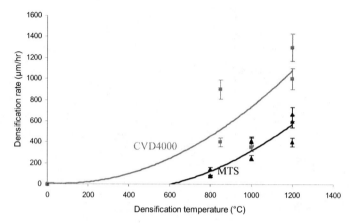

Figure 14. Deposition rate versus temperature for two SiC precursors

CONCLUSION

SiC matrices were deposited by the film-boiling process from two precursors. The MTS allows obtaining dense deposits but containing a carbon excess. On the contrary, the SiC elaborated from CVD4000 has a homogeneous composition. The densification rate is strongly reduced with the film-boiling technique, compared to CVI. The works presented in this paper were achieved on an experimental device to test different expensive precursors in few volumes. The further works will consist in transferring the conclusive experiments on a bigger facility with an inductive heating allowing a wider performs choice to elaborate dense composites and test their physical properties.

ACKNOWLEDGEMENT

The authors wish to thank F. Schuster (Director of Advanced Material Program in CEA) and the Région Centre for their financial support through a PhD grant. Numerous interesting discussions about samples characterizations and results interpretations with D. Demattei and A. M. Lecreurer-Herail (CEA Le Ripault) have been essential to the progress of this study.

REFERENCES

[1]B. C. Bernard, C. Robin-Brosse, J. C. Cavalier, Method of densification of porous substrate by a matrix containing carbon, *European Patent n° 92400075.5*, 23/04/1992.

[2]M. Rosso, Ceramic and metal matrix composites: Routes and properties, *Journal of Materials Processing Technology*, **175**, 364-75 (2006).

[3]R. Naslain, Design, Preparation and properties of non oxide CMCs for application in engines and nuclear reactors: an overview, *Composites Science and Technology*, 64, 155-70 (2004).

[4]T. M. Besmann, R. A. Lowden, D. P. Stinton, Overview of chemical vapor infiltration, *Eds. R. Naslain, J. Lamon, D. Doumeingts*, Proceedings of the 6th European Conference on Composite Materials, September 20-24, 1993, Bordeaux, HT-CMC1, 215-29 (1993).

[5]I. Golecki, R.C. Morris, D. Narasimhan, Method of rapidly densifying a porous structure, *US Patent n°5.348.774* (20/09/1994).

[6]I. Golecki, R. C. Morris, D. Narasimhan, N. Clements, Rapid densification of porous carbon-carbon composites by thermal-gradient chemical vapor infiltration, *Applied Physics Letter*, **66**, 2334-36 (1995).

[7]P. Delhaès, Chemical vapour deposition and infiltration processes of carbon materials, *Carbon*, **40**, 641-57 (2002).

[8]R. Naslain, R. Pailler, X. Bourrat, S. Bertrand, F. Heurtevent, P. Dupel, F. Lamouroux, Synthesis of highly tailored ceramic matrix composites by pressure-pulsed CVI, *Solid* State Ionics, **141-142**, 541-48 (2001).

[9]M. Houdayer, J. Spitz, D. Tran Van, Process for the densification of a porous structure, *US Patent n°4.472.454* (18/09/1984).

[10]P. Delhaes, M. Trinquecoste, A. Derré, D. Rovillain, P. David, Film Boiling Chemical Vapor Infiltration of C/C Composites : Influence of Mass and Thermal Transfers, *Carbon Science*, **4**, 163-67 (2003).

[11]D. Rovillain, M. Trinquecoste, E. Bruneton, A. Derré, P. David, P. Delhaès, Film boiling chemical vapor infiltration: an experimental study on C/C composites materials, *Carbon*, **39**, 1355 -65 (2001).

[12]S. Beaugrand, Etude du procédé de densification rapide par caléfaction pour l'élaboration de composites carbone/carbone, *PhD*, University of Orléans (2000).

[13]K. J. Hüttinger, W. Benzinger, Chemistry and kinetics of Chemical Vapor Infiltration of Pyrocarbon: VI. Mechanical and structural properties of infiltrated carbon fiber felt, *Carbon*, **37**, 1311-22 (1999).

[14]S. Webb, K. Pruess, The use of Fick's Law for modelling trace gas diffusion in porous media, *Transport in Porous Media*, **51**, 327-41 (2003).

[15]A. Serre, J. Blein, P. David, F. Audubert, S. Bonnamy, Study of the ceramic matrix composites densification by film-boiling process, Proceedings ICRACM 2010 (2010).

[16]B.J. Choi, D.R. Kim, Growth of silicon carbide by chemical vapour deposition, *Journal of Materials Science Letters*, **10**, 860-62 (1991).

[17]F. Loumagne, F. Langlais, R. Naslain, Experimental kinetic study of the chemical vapour deposition of SiC-based ceramics from CH_3SiCl_3/H_2 gas precursor, *Journal of Crystal Growth*, **155**, 198-204 (1995).

[18]R. Naslain, R. Pailler, X. Bourrat, S. Bertrand, F. Heurtevent, P. Dupel, F. Lamouroux, Synthesis of highly tailored ceramic matrix composites by pressure-pulsed CVI, *Solid State Ionics*, **141-142**, 541-48 (2001).

[19]M. Placide, Interfaces dans les revêtements de carbure de silicium, *PhD*, University of Bordeaux I, n°3442 (2007).

[20]Y. Awad, S. Allen, M. Davies, A. Gaumond, M. A. El Khakani, R. Smirani, Method for forming a film on a substrate, *US Patent n° 0.129.994 A1* (27/05/2010).

[21]M. Placide, Interfaces dans les revêtements de carbure de silicium, *PhD*, University of Bordeaux I, n°3442 (2007).

[22]F. Kobayashi, K. Ikawa, K. Iwamoto, Formation of carbon-excess SiC from pyrolysis of CH3SiCl3, *Journal of Crystal Growth*, **28**, 395-96 (1975).

[23]J. Heinrich, S. Hemeltjen, G. Marx, Analytics of CVD processes in the deposition of SiC by Methyltrichlorisilane, *Mikrochimica Acta*, **133**, 209-14 (2000).

[24]C. Lu, L. Cheng, C. Zhao, L. Zhang, Y. Xu, Kinetics of chemical vapor deposition of SiC from methyltrichlorosilane and hydrogen, *Applied Surface Science*, **255**, 7495-99 (2009).

PROCESSING OF ULTRAFINE BETA-SILICON CARBIDE POWDER
BY SILICON–CARBON REACTION

S. Sonak, S. Ramanathan and A. K. Suri
Materials Group
Bhabha Atomic Research Centre, Mumbai, 400085, India
sonaks@barc.gov.in

ABSTRACT
Silicon Carbide (SiC) finds applications as a structural ceramic in high temperature nuclear reactors and processing of the components for such applications require sinter-active silicon carbide powders. Synthesis of powders by Silicon, Carbon has been studied. Phase pure powder formation was found to occur in the temperature range of 1673 to 1823K as shown by Differential thermal analysis and X Ray Diffraction studies. The Silicon Carbide powders formed were agglomerated (Mean particle size ~10microns) with faceted dense particles. The dispersion conditions for effective wet grinding were optimized through zeta-potential and rheological studies. A pH of ~8 was found to be optimum for dispersion (Zeta potential ~ -30mV) and the slurry with a solid loading of 1 g/cc (i.e., shear rate independent Newtonian flow). After de-agglomeration for 4 hours in a planetary mill, powders with a mean particle size of ~2microns free of agglomerates with size greater than 10 microns formed. These powders were incorporated with Yttria-Alumina and Aluminium-Boron-Carbon additives and subjected to hot pressing at 2223K under the pressure of 50MPa in protective atmosphere. The sintered bodies upon characterization were alpha modification with a density greater than 96%T.D. exhibiting faceted/lath like microstructure.

INTRODUCTION

Silicon carbide is widely known for its high temperature application under non-oxidative environment due to its stability, high hardness, high thermal conductivity, low density etc. These properties make it an attractive material for high temperature applications in heat exchangers, gas turbines, automobile parts, heating elements, abrasive tools etc.[1].In addition to above, properties such as high irradiation resistance and low cross section for neutron absorption make silicon carbide the a very useful for high temperature applications in nuclear industry. Silicon carbide and silicon carbide composites are actively being researched for the first wall material of the fusion reactor and also for the structural components in upcoming high temperature (Gen IV) reactors[2, 3].

Structural components made out of silicon carbide ceramics are expected to remain hard and resist deformation. The key is to attain high strength by obtaining highly dense sintered product with fine grain size. These components are fabricated using standard ceramic processing technique which involves various stages such as powder synthesis, green body preparation and sintering. Each stage is critical as the flaws introduced in any stage can be detrimental to the performance and the service life of the component. Lot of studies on powder synthesis, green shape forming and its densification behavior has been reported[4-12].

Commercially Silicon carbide is produced by the Acheson process, where carbothermic reduction of silicon dioxide (SiO_2) by carbon is carried out in which temperature of the reaction is about 2273 K[13]. The process is highly energy intensive and α-SiC is the major reaction product along with considerable amount of impurities. Since α-SiC has hexagonal crystal symmetry, it creates difficulty during the process of sintering[14]. The cubic phase of SiC i.e. β-SiC is often desired which undergoes phase transformation to form platelike β-SiC grains during sintering which enhance the fracture toughness by bridging and deflecting the cracks around the elongated α-SiC grains. Many methods could be used to produce β-SiC powders, such as carbo-thermal reduction, sol.gel methods, gas-phase reaction method, self propagation high-temperature synthesis (SHS) etc[4-12]. Powders thus synthesized need to be treated to bring to tailored characteristics to form green bodies with homogeneous microstructure. The as formed product is agglomerated and need to be wet ground under optimum dispersion conditions to obtain fine sinter-active powders. Due to covalent bonded structure of silicon carbide, the diffusion of atomic species are restricted by high activation barrier which makes the sintering a difficult process in these systems. Hence to enhance the process of densification, sintering of the system is generally carried out using hot pressing technique incorporating additives that can form liquid phase at lower temperature (~2173K) .Various sintering additives such as aluminium, boron, carbon, alumina-yttria, aluminium nitride etc for silicon carbide and their roles were investigated over the last three decades[15-18].

In the present study, conditions for formation of beta-silicon carbide (β-SiC) by the reaction of Si and C was optimized using thermal analysis and X Ray diffraction. As the product formed was highly agglomerated, dispersion conditions for effective wet grinding to form into fine powder (deagglomeration) has been studied using zeta-potential, rheological and particle size distribution analysis. Phase pure and fine powders thus produced were incorporated with and without additives and hot pressed to get sintered pellets. The results obtained are presented and discussed.

EXPERIMENTAL

The stoichiometric amount of silicon powder (Si-98.5 % ,4.3μm median dia., supplied by M/S SRL Pvt. Ltd., Mumbai, India) and petroleum coke (C-99.4%,13.9 μm median dia., supplied by M/S Assam carbon, India) having agglomerate size ~ 5-8microns were planetarily wet mixed, dried, granulated and compacted to reaction specimens. Thermal analysis of the reaction mass was carried out to identify the temperature range to be used for reaction. Based on this study, the reaction compacts were heated in a super-kanthal furnace in argon atmosphere in the temperature range of 1673 to 1873K for varying time. The reacted specimens were subjected to XRD studies for progress of reaction. The products were de-agglomerated by dry grinding and the dispersion conditions for wet grinding were optimized by zeta-potential and viscosity of slurries with varying solid content. Wet grinding was carried out in a planetary ball mill using tungsten carbide pot and ball. The agglomerate size was determined using particle size distribution by laser light scattering technique. The fine powders formed were incorporated with yttria- alumina and boron-carbon-aluminum mixtures. Sintering was carried out in a hot press at 2223K at a pressure 50MPa. The sintered specimens were characterized for phase change by XRD, density by Archimedes principle and microstructure by SEM.

RESULTS AND DISCUSSIONS

The reactants were wet mixed to ensure homogeneity in mixing which is essential to bring about the reaction. The thermal analysis data of the silicon-carbon reaction is shown in fig.1. There was no heat effect until 1673K above which there was setting in of an endotherm that was merged with an exotherm. This can be attributed to the melting of silicon (endotherm) followed by a reaction of carbon with liquid silicon (exotherm). The overlap of both these heat effects clearly brings out the enhancement in the reactivity in liquid-solid reaction compared to solid-solid reaction. This was noticed in XRD data as shown in Fig.2. It is noteworthy that the product is beta form of SiC which has been reported to be sinteractive one[19]. Even though this process is reported to be a typical self propagating high temperature (SHS) reaction, the heat of this reaction is low so that it requires heating to a temperature of the melting to enhance the kinetics of the process. It is interesting to note by a keen observation, that a little excess amount of carbon was essential for completion of reaction in larger batch size (50g) which was achieved by using 5 to 10 weight% excess carbon. The excess carbon left behind in the product could be selectively eliminated by heating the reaction product at 1173K in air. This could be attributed to the blocking of silicon by the formed SiC product from reaching the carbon. Due to enhanced kinetics of reaction between liquid silicon and carbon, the exotherm was less intense than otherwise. A study of the effect of reaction temperature for a given time of about 1hour in the range of 1723 to 1823K indicated that phase pure beta silicon carbide form only at 1823K (Fig.2). It is obvious the reaction in sets in around the melting point of silicon in increasing the temperature of the reaction to 1823K reduced the time for complete conversion to 1.5 hrs due to enhanced kinetics of the process and phase purity was confirmed by the XRD studies.

The product was powdered and was found to be highly agglomerated. As formation of green bodies with homogeneous and fine microstructure requires de-agglomeration, this behavior was studied by dry and wet grinding in planetary ball mill using tungsten carbide pot. The importance of dispersion conditions on the effective wet grinding is emphasized in the study by Houivet et al.[20]. For effective wet grinding, the suspended agglomerates need to be well dispersed which can be achieved by imparting charge on the surface of the particle. The charging results in columbic repulsion between particles thereby stabilizing the suspension. Also the slurry solid content should be such that it should exhibit Newtonian flow behavior under the milling conditions. Flocculates are reported to be present in concentrated slurries that results in pseudoplasticity which drastically reduces the state of dispersion. The optimum condition for charging the particles can be obtained through the zeta-potential measurement as a function of pH of the medium. The result shown in fig.3. exhibit that the zeta potential is maximum numerically around a pH of 8. The viscosity variation with shear rate plot of the slurries with varying solid content as shown in fig.4 exhibited Newtonian flow behavior for slurries up to a concentration of 1g/ml. Hence, slurries with this concentration at pH 8 were used for grinding. The particle size distribution with time of grinding is shown in fig.6 and it is obvious the dry ground powder is exhibits a size of 12microns with maximum size up to 100microns. With progress of wet grinding the size could be reduced to about 1 micron with a narrower size distribution that is needed for forming green bodies with homogeneous and fine pore structure. The study on densification behavior of the powders by hot pressing exhibited the need for additives (Table.1.).

Table.1. Role of additive on densification behavior of SiC

Sr. No	Additives	Density
1	No additives	66.23%
2	Yttria-Alumina	86 %
3	Aluminium-Boron-Carbon	96%

Even though addition of yttria-alumina with the eutectic composition has been reported to yield densification, it could result in bodies with only 86% T.D. However, the addition improves densification compared to the powder without additive. However, the addition of aluminum boron carbon has been found to yield the maximum density of the order of 96% TD. It is interesting to note that the hot pressed pellets could be easily removed from the die punch as there was no interface reaction. However, in case of yttria-alumina addition, there was reactivity that made removal of the pellet from the die a bit difficult which suggests the need for suppressing of the reactivity through methods such as applying non reactive coating or additional additive. The typical microstructure of the fractured samples sintered with-out and with additives are shown in Fig.6. It is obvious that the pore fraction decreased as expected with improved densification. The grains exhibited equiaxed morphology with plate like features. The XRD study of the densified samples exhibited transformation of beta to alpha phase of the silicon carbide (Fig.7) which also is expected to enhance densification in solid state sintering. However, additives used in this study have been reported to improve densification through the liquid phase formation[21].

CONCLUSIONS

1. Sinteractive beta silicon carbide powders with desired characteristics were synthesized.
2. From X-ray-diffraction studies, a complete conversion of silicon and carbon into silicon carbide has been achieved by optimizing the reaction temperature and the time.
3. The dispersion conditions for effective wet grinding to form into fine powder has been optimised using zeta-potential, rheological and particle size distribution analysis.
4. The combination of Al, B, C acted as an effective sintering aid in SiC as compared to yttria and alumina. Hot pressing with Al, B, C additives resulted in near theoretically dense SiC samples.
5. Liquid phase sintering is the prominent mechanism of the sintering.

ACKNOWLEDGMENT
Authors would like to thank Dr. Abhijit Ghosh, Mr. M.R.Gonal , Dr.A.K.Sahoo, Mr. S. Syambabu and S.Koley for their help during the experimentations.

REFERENCES

1. R.W.Cahn, P.Hassen & E.J.Kramer, Material Science and Technology, Vol. 17A, (1996) pp.129-150.
2. L.L. Snead , R.H. Jones, A. Kohyama, P. Fenici ,Status of silicon carbide composites for fusion , J. Nucl. Mater. 233-237 (1996) 26-36.
3. H.L. Heinisch , L.R. Greenwood, W.J. Weber, R.E. Williford, Displacement damage in silicon carbide irradiated in fission reactors, J. Nucl. Mater. 327 (2004) 175–181.
4. Yun Yang, Zhi-Ming Lin, Jiang-Tao Li, Synthesis of SiC by silicon and carbon combustion in air, J. the Eur. Ceram. Soc. 29 (2009) 175–180.
5. S. Larpkiattaworn, P.Ngernchuklin, W.Khongwong, N.Pankurddee, S. Wada, The influence of reaction parameters on the free Si and C contents in the synthesis of nano-sized SiC, Ceram. Int. 32 (2006) 899–904.
6. L.N.Satapathy, P.D.Ramesh , D. Agrawal , R.Roy, Microwave synthesis of phase-pure, fine silicon carbide powder, Mater. Res. Bull. 40 (2005) 1871–1882.
7. R.M. Ayral, F. Rouessac, N.Massoni, Combustion synthesis of silicon carbide assisted by a magnesium plus polytetrafluoroethylene mixture, Mater.Res. Bull. 44 (2009) 2134–2138
8. Y. Okabe, J.Hojo and A.Kato, Formation of fine Silicon Carbide powders by a vapor Phase Method, J. Less-Common Met. 68 (1979) 29 – 41.
9. Z.Yermekova, Z.Mansurov, A.Mukasyan, Influence of precursor morphology on the microstructure of silicon carbide nanopowder produced by combustion synthesis, Ceram. Int. 36 (2010) 2297–2305.
10. A.Can, M.Herrmann D.S.McLachlan, I.Sigalas, J. Adler, Densification of liquid phase sintered silicon carbide, J. Eur. Ceram. Soc. 26 (2006) 1707–1713.
11. Y.Zhou, H. Tanaka, S. Otani, and Y. Bando, Low-Temperature Pressureless Sintering of α-SiC with Al$_4$C$_3$–B$_4$C–C Additions ,J. Am.Ceram. Soc.82 (1999) 1959–64.
12. M.A.Mulla, V.D.Krstic, Pressureless sintering of beta-SiC with Al$_2$O$_3$ additions, J. Mater.Sci.29 (1994) 934-938

13. V.D. Krstic, Production of fine, high-purity beta silicon carbide, J. Am. Ceram. Soc.75 (1992) 170.
14. A. M. Kueck,C. Lutgard , De Jonghe , Two-stage sintering inhibits abnormal grain growth during to transformation in SiC, J. Eur. Ceram. Soc. 28 (2008) 2259–2264.
15. Y.Kim, W.Mitomo, M. Zhan, Microstructure control of liquid-phase sintered β-SiC by seeding, J.Mat. Sci.Lett. 20 (2001) 2217.
16. K. Biswas, Liquid Phase Sintering of SiC-Ceramics, Mater. Chem. Phys. 67 (2001)1578
17. K.Biswas, Solid State Sintering of SiC-Ceramics, Mater. Sci. Forum Vol. 624 (2009) 91-108.
18. S.Prochazaka, Special Ceramics, Ed. P. Popper, Vol. 6 171-84(1975)
19. G.C.Wei, C.R. Kennedy, L.A.Harris, Synthesis of sinterable SiC powders by carbothermic reduction of gelderived precursors and pyrolysis of polycarbosilane, J.Am.Ceram. Soc. Bull. 63 (1984) 1054.
20. D.Houivet, J.E.Fallah and J.M. Haussonne Dispersion and grinding of oxide powders into an aqueous slurry, J.Am. Ceram. Soc. 85 (2002) 321.
21. A.Ghosh, A.K.Gulnar, R.K.Fotedar, G. K.Dey, A.K. Suri, Microstructural Studies of Hot Pressed Silicon Carbide Ceramic, Ceramics in nuclear engineering, Pages: 113–122, 2010 Published Online : 14 JAN 2010, DOI: 10.1002/9780470584002.ch10

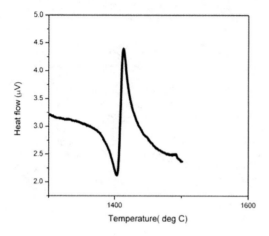

Figure 1 : Thermograph of the silicon – carbon reaction

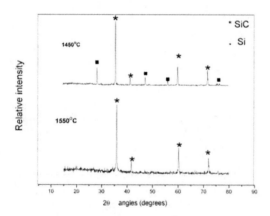

Figure 2 : XRD patterns of the partially and fully reacted silicon carbide powders

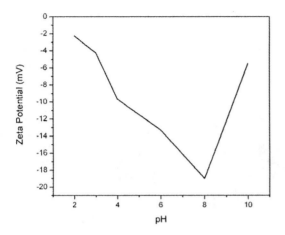

Figure 3 : Zeta potential variation of SiC dispersion against pH

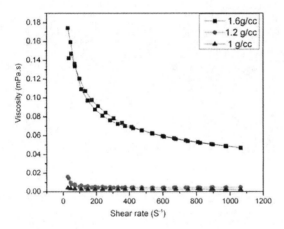

Figure 4: Viscosity variation with shear rate plot of the slurries with varying solid content

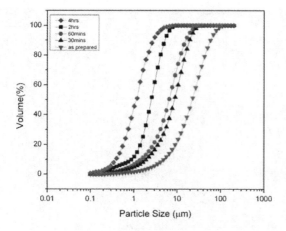

Figure 5 : The particle size distribution with time of grinding

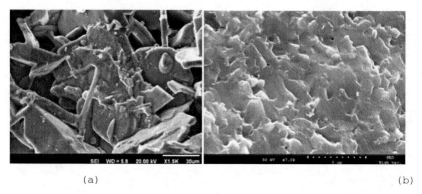

(a)   (b)

Figure 6   The typical microstructure of the fractured samples sintered a) with-out and b) with additives

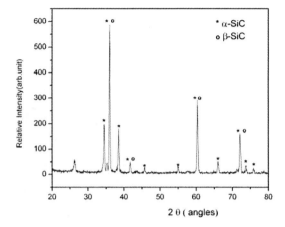

Figure 7 : The XRD pattern of the densified samples exhibiting transformation of beta to alpha phase of the silicon carbide

CHARACTERIZATION OF FAILURE BEHAVIOR OF SILICON CARBIDE COMPOSITES BY ACOUSTIC EMISSION

Takashi Nozawa, Kazumi Ozawa, Hiroyasu Tanigawa
Japan Atomic Energy Agency,
2-166 Omotedate, Obuchi, Rokkasho, Aomori 039-3212, Japan

ABSTRACT

A silicon carbide fiber reinforced silicon carbide matrix (SiC/SiC) composite is a promising candidate material for various nuclear applications such as a fusion blanket. Due to the inherent brittle-like failure, identifying failure scenario with consideration of reliability and reproducibility of composite's characteristics is undoubtedly important to develop design codes specialized for composites. This study aims to identify failure behavior of SiC/SiC composites by varied test modes. For this purpose, acoustic emission (AE) was applied to detect composites' failure. The tensile, compressive and in-plane shear tests were conducted for plain-weave (P/W) chemical-vapor-infiltration (CVI) SiC/SiC composites. Various loading angles were applied to discuss an anisotropic issue. AE results distinguished damage accumulation processes in axial and off-axial loading cases. Specifically, test results indicated a clear difference of damage density between tensile and compressive tests. This study also classified the characteristic failure modes by separately discussing localized variations of power within a time series by wavelet analysis.

INTRODUCTION

A SiC/SiC composite is a promising candidate material for wide spread advanced nuclear applications such as blanket structural materials for a fusion demonstration power reactor (DEMO) [1, 2] and a control rod sheath of the Very High-Temperature Reactor (VHTR) [2] due to the excellent irradiation resistance, low after-heat and low-induced radioactivity as well as recognized several merits as engineering ceramics: high-strength and chemical inertness at high-temperatures.

For the engineering design of the practical applications, identifying the failure scenario and lifetime under practical environments is undoubtedly important. For instance, considering the fusion DEMO environments, the high heat flux induced by the fusion plasma gives a steep thermal gradient inside the material. This thermal gradient subsequently gives differential swelling under irradiation, as well as differential thermal expansion, resulting in the complex stress condition [3]. Mechanical integrity under such a non-uniform stress condition, therefore, needs to be ensured for practical applications even for the functional structure applications.

In general, composites fail in a quasi-ductile manner with permanent damage accumulation. The failure behavior of the quasi-ductile composites is therefore quite different from those of brittle ceramics and ductile metals, and understanding the failure process is indispensable. Specifically, the failure behavior of the nuclear-grade SiC/SiC composites, which are generally defined as SiC/SiC composites composed of highly-crystalline and near-stoichiometric SiC matrix (e.g., CVI-SiC) and fibers (e.g., Hi-Nicalon Type-S and Tyranno-SA), needs to be identified. From the recent reports by the authors [4, 5], the failure behavior of this class of composites was evaluated by varied failure modes, with consideration of strength anisotropy. In this study, it was found that SiC/SiC composites show the axial anisotropy, i.e., significant strength gap between the tensile strength and the compressive strength. Also, the in-plane shear fracture behavior depends on the test methodologies. From the micromechanics approach to evaluate the F/M interfacial strength, the significant contribution of the strong interfacial debonding strength and friction stress to bulk strength was identified [6]. These behaviors are a little bit different from the conventional SiC/SiC composites with comparably weak F/M interfacial strength.

The acoustic emission technique has been employed to monitor damage accumulation process of the composites early in the stage of composites' R&D and will become more common as in-situ damage monitoring tool. There are several attempts to classify the AE parameters with the characteristic failure modes [7]. Using AE data, the modeling technique to predict stress-strain behavior of woven SiC/SiC composites was developed [8].

This paper aims to evaluate detailed failure behavior of SiC/SiC composites by various modes such as in-plane tensile/compressive, and in-plane shear coupled with the AE measurements for the latest nuclear-grade SiC/SiC composites. Specifically, the wavelet technique, one of the potential powerful analytical tools, was applied to monitor the time-dependent behavior of the AE event and applicability of the technique was discussed.

EXPERIMENTAL

Material

A P/W CVI-SiC/SiC composite was used (Hyper-Therm High-Temperatures Composites, Inc., Huntington Beach, CA, USA). Highly-crystalline and near-stoichiometric Tyranno-SA3 SiC fibers (Ube Industries, Ltd., Ube, Japan) were used as reinforcements. A single layer of pyrolytic carbon (~150 nm) was formed on the fiber surface as the F/M interphase. The fiber volume fraction and the porosity were ~30% and ~18%, respectively.

Mechanical Tests

Tensile and compressive tests were conducted at room temperature with varied loading directions based on the guidelines of ASTM C1275 and C1358, respectively. A rectangular face-loaded specimen with a size of 40L×4W×1.3~2.0T mm (a gauge length of 15 mm) was applied to both tests. Aluminum tabs were bonded on the gripping ends of the specimen to prevent from damage at the gripping section during the test. Cyclic loading tests were conducted to monitor damage accumulation process during unloading/reloading sequences. Tensile and compressive strains were measured by a couple of strain gauges with a gauge length of 5 mm bonded on the center of the specimen surfaces and an average reading was used in calculation as a representative strain of the composites. The crosshead displacement rate was 0.5 mm/min.

In-plane shear properties were evaluated by the Iosipescu method (ASTM C1292) at room-temperature. A schematic of the Iosipescu test is shown in Figure 1. The miniature specimen was used. The shear strain was measured by a pair of rosette-type (two perpendicular directions) strain gauges with a gauge length of 2 mm bonded on the specimen surfaces. The crosshead displacement rate was 0.5 mm/min. One of the fiber longitudinal directions was parallel to the loading direction.

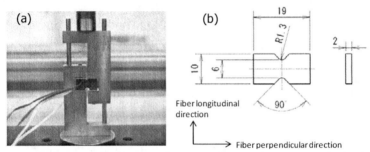

Figure 1. (a) A photo image of miniature Iosipescu test fixture and (b) a schematic drawing of the miniature Iosipescu test specimen (mm in unit).

To monitor AE signal during fracture of SiC itself, the three-point bend test was conducted for chemical-vapor-deposition (CVD) SiC at room-temperature. The rectangular bend specimen with a size of 20L×2W×1T mm was used. The support span was 16 mm. The crosshead displacement rate was 0.5 mm/min.

Fracture surfaces of the tested SiC/SiC composites were observed by the field emission scanning electron microscopy (FE-SEM).

Acoustic Emission Measurements

Acoustic emission events were monitored during the mechanical tests using a commercial measurement system (AMSY-5, Vallen Systeme GmbH, Germany). In this study, a pair of AE sensors with the resonance frequency of 100 kHz (AE104A, Vallen Systeme GmbH, Germany) was attached on the grip surfaces because of the limited area of the specimen. While, the AE sensor was bonded directly on the tensile surface of a CVD-SiC bend specimen. The measured AE signals were amplified by the pre-amplifier of 34 db. The band pass filter of 25-850 kHz was set.

The obtained data were analyzed by the standard software provided with the system (VisualAE and VisualTR). Additionally, wavelet analysis was applied to evaluate localized variations of power within a time series. By decomposing a time series into time–frequency space, one is able to determine both the dominant modes of variability and how those modes vary in time. For this analysis, Gabor wavelet which is based on the Gaussian function was utilized [9].

RESULTS

Tensile Test

Figure 2 shows typical tensile stress-strain curves of P/W CVI-SiC/SiC composites by either 0-deg. or 45-deg. loading. In these figures, the cumulative AE energy is plotted together to monitor the damage accumulation process. For both 0-deg. and 45-deg. loading cases, the tensile stress-strain curves show non-linear damage accumulation beyond the initial linear elastic region. The transition stresses (i.e., the proportional limit stresses (PLS)) are 139 MPa and 132 MPa for 0-deg. and 45-deg. cases, respectively, showing no significant anisotropy of the PLS. On the other hand, the fracture strength depends on the loading angle (254 MPa for 0-deg. vs. 183 MPa for 45-deg.). The AE energy was induced prior to the proportional limit. For the 0-deg. case, the cumulative AE energy >1000 eu was measured at ~100 MPa (~70% of PLS) and exponentially increased with increasing applied stress. Note that 1 eu is 1 nVs. It was also found that the rate of increase of the AE energy was changed drastically with increasing stress. Noteworthy the transition of this rate change corresponds to the PLS. The same is true for the 45-deg. tensile loading case. There was no AE event detected during unloading/reloading sequences for both cases because of the Kaiser effect, which is common for many types of composites.

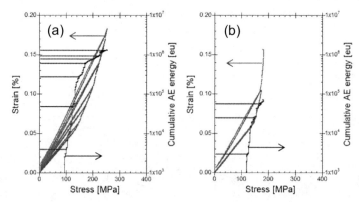

Figure 2. Typical tensile strain vs. stress curves of P/W CVI-SiC/SiC composites: (a) 0-deg. and (b) 45-deg. tensile loading cases. Cumulative AE energy during the test is plotted together.

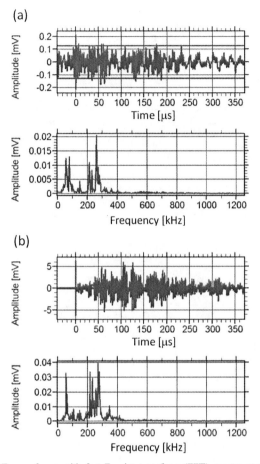

Figure 3. Typical AE waveforms with fast Fourier transform (FFT) spectra at the proportional limit stress for P/W CVI-SiC/SiC composites: (a) 0-deg. and (b) 45-deg. tensile loading cases.

Figure 3 shows typical AE waveforms with fast Fourier transform (FFT) spectrum at the PLS. From the FFT spectrum, the major peaks of characteristic frequencies were obtained in the ranges of <100, 100-150, 200-300 and 300-400 kHz. There seems no significant difference in this distribution between the 0-deg. and the 45-deg. tensile loading cases.

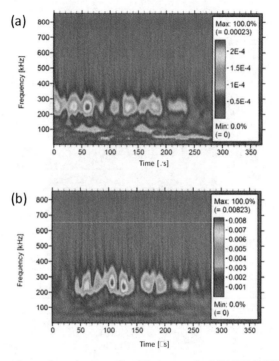

Figure 4. Typical wavelet results at the proportional limit stress for P/W CVI-SiC/SiC composites: (a) 0-deg. and (b) 45-deg. tensile loading cases.

Figure 4 shows wavelet results from the same waveform data in Fig. 3. The figure shows the frequency vs. time with intensity. For both 0-deg. and 45-deg. cases, the strong peak of the characteristic frequency was obtained at ~260 kHz. From the FFT spectrum of Fig. 3, several peaks were obtained in the range of 200–300 kHz but, by plotting the time dependence (Fig. 4), it was found that the clusters of peaks with the peak frequency of ~260 kHz were periodically shown. Besides, from the figures, several weak peaks were obtained at ~50, ~100 and ~130 kHz.

Figure 5 shows typical tensile fracture surfaces of P/W CVI-SiC/SiC composites by either 0- or 45-deg. loading. From the figures, the quasi-ductile fracture pattern is apparent for both loading cases. Specifically, as generally observed in Tyranno-SA3 composites, very short fiber pullouts were clearly shown due primarily to very strong interfacial bonding strength and friction stress. In contrast, the longer fiber pullouts appeared to be obtained for the 45-deg. tensile loading case. The reasonable explanation is that the shear failure mode is more dominant in the off-axial loading case.

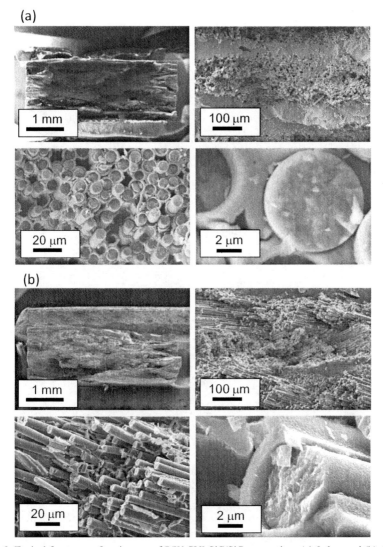

Figure 5. Typical fracture surface images of P/W CVI-SiC/SiC composites: (a) 0-deg. and (b) 45-deg. tensile loading cases.

Compressive Test

Figure 6 shows the typical axial (0-deg.) compressive failure behavior of P/W CVI-SiC/SiC composites. The composites failed in a brittle manner by compressive loading, while the composites failed in a quasi-ductile manner by tensile loading. The higher PLS of ~372 MPa was obtained compared with the tensile loading cases. In contrast, the fracture strength is almost equivalent to the PLS. The cumulative AE energy was induced at the low stress level even though the stress-strain curve still shows linearity. Significant increase of the cumulative AE energy was observed at the stress level of 50% of the PLS.

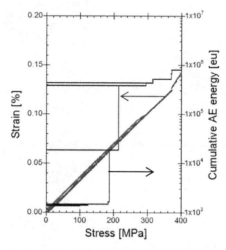

Figure 6. The typical compressive strain vs. stress curve of P/W CVI-SiC/SiC composites. Cumulative AE energy during the test is plotted together.

Figure 7 shows a typical AE waveform with a FFT spectrum at the PLS by 0-deg. compressive loading for P/W CVI-SiC/SiC composites. A strong peak of ~350 kHz was obtained as well as a peak of ~260 kHz, which was also obtained for the tensile loading cases.

Figure 8 shows a typical wavelet result at the proportional limit stress by 0-deg. compressive loading for P/W CVI-SiC/SiC composites. The shape of the cluster was slightly prolonged to the higher frequency region. And intervals of the appearance of the clusters were shortened.

Figure 9 shows typical compressive fracture surface images of P/W CVI-SiC/SiC composites. The composites show less fiber pullouts by compressive loading and most of the F/M interfaces were intact after the test, implying no significant matrix cracking at the interface. The primary mechanism of failure is alternatively rapid matrix break over the cross-section of the specimen at the maximum stress.

Although the stress vs. strain curve shows brittle fracture pattern, with very limited numbers of fiber pullouts, this fracture pattern is not exactly same with the brittle fracture as generally observed in ceramics. We often call this brittle-like fracture.

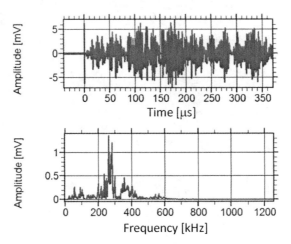

Figure 7. Typical AE waveform with a fast Fourier transform (FFT) spectrum at the proportional limit stress by 0-deg. compressive loading for P/W CVI-SiC/SiC composites.

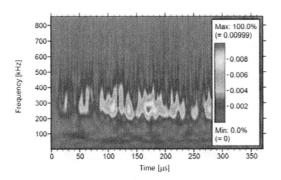

Figure 8. Typical wavelet result at the proportional limit stress by 0-deg. compressive loading for P/W CVI-SiC/SiC composites.

Figure 9. Typical compressive fracture surface images of P/W CVI-SiC/SiC composites.

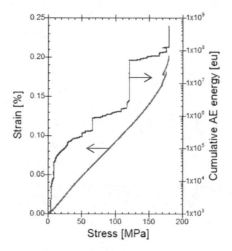

Figure 10. The typical in-plane shear strain vs. stress curve of P/W CVI-SiC/SiC composites. Cumulative AE energy during the test is plotted together.

In-Plane Shear Test

  Figure 10 shows in-plane shear fracture behavior by the Iosipescu test for P/W CVI-SiC/SiC composites. The composites show initial linearity followed by quasi-ductility. Although there was a clear initial linear segment, the AE could detect significant damage accumulation from the beginning of the test. The cumulative AE energy continuously increases with increasing stress until composite's fracture. The final cumulative AE energy was two or three order higher than those of tensile and compressive cases.

  Figure 11 shows typical wavelet results of the Iosipescu test for P/W CVI-SiC/SiC composites at stress levels of 25 MPa and 150 MPa. The latter stress level corresponds to the PLS. In both cases, strong continuous peak of ~75 kHz was obtained. Also periodic peaks at ~130, 200 and 260 kHz were obtained. Although it is hard to explicitly distinguish the difference between these two plots, it seems that the peak of ~260 kHz was more dominant at 150 MPa.

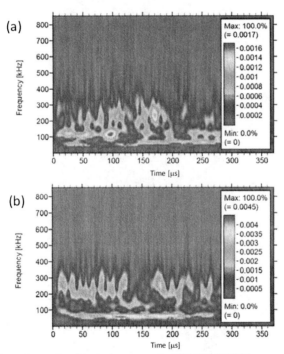

Figure 11. Typical wavelet results of the Iosipescu test for P/W CVI-SiC/SiC composites: (a) at 25 MPa and (b) at 150 MPa.

Bend Test for CVD-SiC

Figure 12 shows the typical wavelet result at the fracture stress by 3-point bend loading for CVD-SiC. Strong clusters of peaks at ~260 kHz were clearly obtained, as well as a continuous peak at ~75 kHz.

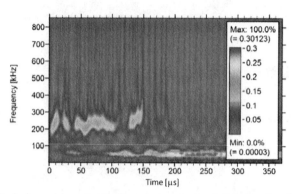

Figure 12. Typical wavelet result at the fracture stress by 3-point bend loading for CVD-SiC.

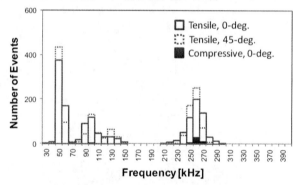

Figure 13. Number of AE events vs. frequencies.

DISCUSSION

To evaluate detailed damage accumulation process of the nuclear-grade SiC/SiC composites, the distribution of the characteristic frequencies was summarized by the cumulative number of the AE events appeared in each frequency for the axial and off-axial tensile tests and the axial compressive tests (Figure 13). The result specified the four characteristic peaks at ~50, ~100, ~130 and ~260 kHz. As clearly seen in Figure 14, the peak at ~50 kHz was derived from the noise, which seems primarily attributed to the test setup. At ~100 kHz, the resonance of the AE sensor may affect the results. So the remaining two frequencies are typical for the composites' failure. The peak at ~260 kHz was common for all cases, while the peak at ~130 kHz was typical only for the tensile cases regardless of the loading directions.

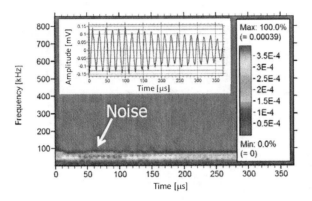

Figure 14. Typical wavelet result by system noise.

Table I. Summary of AE measurements

Material	Test mode	Ave. # of total AE events	Ave. # of total AE counts	Ave. count rate	Ave. total AE energy [eu]
P/W CVI SiC/SiC composites	Tensile, 0-deg.	864	53975	45	6.22×10^{6}
	Comp., 0-deg.	53	6014	113	9.00×10^{5}
	Tensile, 45-deg.	523	23062	44	5.43×10^{7}
	In-plane shear	3812	162095	42	6.62×10^{8}
CVD-SiC	Bend	16	1367	124	3.57×10^{8}

With a good agreement with the result of CVD-SiC, it should be reasonable that the peak at ~260 kHz indicates matrix cracking. However, due to the fact that the cumulative number of AE events for the compressive case was quite lower than that for the tensile case, the detailed discussion is necessary to conclude this. The results of the AE measurements are summarized in Table I. From the table, compared with the tensile and in-plane shear cases, less average numbers of total AE events were obtained in compressive tests for composites and bend tests for CVD-SiC. However, the average total AE energy was not so deviated. This is because the average count rates for the compressive and bend test cases were about three times higher than those for tensile and in-plane shear cases. The AE count depends on the intensity of the AE waveform and the intensity of the AE waveform can be increased by several waveforms co-existed. This indicates that the crack density in one AE event is comparably high for the composites' compressive case and the ceramics' bend case. This is visually apparent from the wavelet results. The pulsed clusters of AE signals were obtained. For CVD-SiC, the main crack propagated through the entire specimen cross-section rapidly. In this period, very large new crack surface can be induced by one crack, releasing very high energy. In contrast, the compressive fracture surface suggested the matrix break as a primary failure mechanism. Similar to the CVD-SiC case, the high energy was released by a large surface crack initiation for the compressive loading case. On the other hand, for tensile and in-plane shear cases, failure of composites are more dominant by the micro matrix cracking, giving a comparably small new crack surface for each crack. Under this situation, one AE event has small portion of energy. However, these matrix cracks are easily induced during the test and eventually the total number of the cracks is increased. By increasing total number of the matrix cracks, the total AE energy can eventually be increased to the equivalent level for the compressive test case.

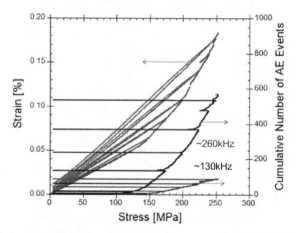

Figure 15. Cumulative numbers of AE event vs. tensile stress curves for P/W CVI-SiC/SiC composites. Tensile strain vs. stress curve is also plotted together. The figure shows the cases of two characteristic frequencies (~130 and ~260 kHz).

Figure 15 shows cumulative numbers of AE events at frequencies of ~130 and ~260 kHz with respect to the tensile stress, coupled with a typical tensile stress vs. strain curve by the axial tensile loading for P/W CVI-SiC/SiC composites. From the figure, it was found that the failure classified by the frequency of ~260 kHz, i.e., matrix cracking, initiates prior the proportional limit, while the failure classified by the frequency of ~130 kHz initiates just beyond the proportional limit. We can simply assume that the latter indicates the initiation of fiber sliding. Due to very rough surface of the highly-crystalline Tyranno-SA3 SiC fibers, the friction stress at the F/M interface is reportedly high. Even though the matrix crack is induced in the SiC matrix, due to the strong friction, the overall composites' stress vs. strain response becomes linear like elasticity with no fiber sliding prior to the proportional limit. Beyond the proportional limit, the fiber can slide, showing a non-linearity. The fact that the peak at ~130 kHz was primarily observed for tensile and in-plane shear test results may support this hypothesis. For the compressive case, the F/M interface was intact.

The applied Tyranno-SA3 fibers are nearly equivalent to CVD-SiC in fracture manner and therefore it is speculated that the characteristic frequency to associate with the fiber break is overlapped in the wide peak range in 200~300 kHz. Further evaluation is necessary to confirm this. At this moment, it is unclear for the peaks at ~200 kHz for in-plane shear tests and at ~350 kHz for mainly compressive tests. They are apparently exists but very minor from the point of the total number of appearance.

Damage accumulation at the very low stress level for compressive and in-plane shear tests is somehow questionable. One possible explanation for such unexpected damage initiation is misalignment of the specimen, which can easily induce the damage. Otherwise buckling would be a potential concern for the compressive loading test cases including Iosipescu tests. Identification of the failure origins using two AE sensors will be helpful to answer them and this will be discussed in the future work.

CONCLUSIONS

Failure evaluation of composites is a critical part to establish design codes and operation scheme and therefore this study aims to evaluate the failure behavior of SiC/SiC composites by tensile/compressive/in-plane shear methods with AE monitoring. Specifically Wavelet analysis was applied to associate characteristic frequencies with individual failure modes. For Tyranno-SA composites, the failure initiation was detected prior the proportional limit stress by the AE technique. No sliding upon matrix cracking was anticipated below the PLS due to strong interfacial shear properties by utilizing rough Tyranno-SA fibers. The types of failure were preliminarily assigned by key characteristic frequencies but further evaluation is necessary to confirm them. Finally, it was implicated that the AE technique with Wavelet analysis becomes a powerful tool to monitor failure behavior of composites, although further optimization is required.

ACKNOWLEDGEMENTS

This paper has been prepared as an account of work assigned to the Japanese Implementing Agency under the Procurement Number IFERC-T1PA01-JA and IFERC-T1PA02-JA within the "Broader Approach Agreement" between the Government of Japan and the European Atomic Energy Community.

REFERENCES

[1]Y. Katoh, L. L. Snead, C. H. Henager, Jr., A. Hasegawa, A. Kohyama, B. Riccardi, and H. Hegeman, Current status and critical issues for development of SiC composites for fusion applications, Journal of Nuclear Materials, **367-370**, 659-71 (2007).

[2]L. L. Snead, T. Nozawa, M. Ferraris, Y. Katoh, R. Shinavski, and M. Sawan, Silicon Carbide Composites as Fusion Power Reactor Structural Materials, *Journal of Nuclear Materials*, **417**, 330-39 (2011).

[3]S. Smolentsev, M. Abdou, N. B. Morley, M. Sawan, S. Malang, and C. Wong, Numerical analysis of MHD flow and heat transfer in a poloidal channel of the DCLL blanket with a SiC_f/SiC flow channel insert, *Fusion Engineering and Design*, **81**, 549-53 (2006).

[4]T. Nozawa, Y. Choi, T. Hinoki, H. Kishimoto, A. Kohyama, and H. Tanigawa, Tensile, Compressive and In-Plane/Inter-Laminar Shear Failure Behavior of CVI- and NITE-SiC/SiC Composites, *IOP Conference Series: Materials Science and Engineering*, **18**, 162011 (2011).

[5]T. Nozawa, K. Ozawa, Y. B. Choi, A. Kohyama, and H. Tanigawa, Determination and prediction of axial/off-axial mechanical properties of SiC/SiC composites, *Fusion Engineering and Design*, (2012) in press.

[6]T. Nozawa, Y. Katoh, and L. L. Snead, The effect of neutron irradiation on the fiber/matrix interphase of silicon carbide composites, *Journal of Nuclear Materials*, **384**, 195-211 (2009).

[7]G. N. Morscher, Modal acoustic emission of damage accumulation in a woven SiC/SiC composites, *Composites Science and Technology*, **59**, 687-97 (1999).

[8]G. N. Morscher, M. Singh, J. Douglas Kiser, M. Freedman, and R. Bhatt, Modeling stress-dependent matrix cracking and stress-strain behavior in 2D woven SiC fiber reinforced CVI SiC composites, *Composites Science and Technology*, **67**, 1009-17 (2007).

[9]H. Suzuki, T. Kinjo, Y. Hayashi, M. Takemoto, K. Ono, Wavelet transform of acoustic emission signals, *Journal of Acoustic Emission*, **14**, 69-84 (1996).

RECESSION OF SILICON CARBIDE IN STEAM UNDER NUCLEAR PLANT LOCA CONDITIONS UP TO 1400 °C.

Greg Markham – Ceramic Tubular Products LLC
Rodney Hall – Ceramic Tubular Products LLC
Herbert Feinroth – Ceramic Tubular Products LLC

ABSTRACT

Ceramic Tubular Products is developing a multilayered silicon carbide tube for use as fuel cladding in commercial nuclear reactors. In this project, we exposed a number of alpha and beta phase SiC monolith specimens to the type of steam conditions that would exist in a commercial water reactor during a design basis LOCA (Loss of Coolant Accident). We also exposed specimens at 1400°C, which is 200°C above the conditions allowed by regulations for a design basis LOCA, and for times up to 8 hours, substantially exceeding the times allowed under current regulatory criteria for licensing of zircaloy clad fuel. The measured recession, and calculated hydrogen gas released during this exposure was compared with that calculated for zircaloy cladding under the same conditions.

For example, for a 4 hour exposure test at 1400°C, the measured recision (loss of clad wall thickness) for SiC, adjusted for actual 17 x 17 LWR fuel clad dimensions, was 0.07%, as compared with 42.1% recession calculated for zircaloy. The data thus demonstrates that use of this cladding in commercial LWRs instead of zircaloy will substantially increase the safety margin of LWRs when subject to Loss of Coolant accidents such as recently occurred at Fukushima.

This work was sponsored by the US Department of Energy under Small Business Innovative Research Grant DE-SC0004225.

INTRODUCTION

The use of silicon carbide in commercial nuclear reactors, particularly as cladding for advanced fuels is postulated to provide substantial safety and economic benefits.[1] Zirconium alloy cladding materials that are presently used in light water reactors present limitations due to their loss of strength during reactor flow transients and other overheating accidents. In addition, the potential for exothermic metal water reactions at design base loss of coolant accident conditions place further limitations. Ceramic Tubular Products is currently developing an advanced ceramic composite tubing meant to overcome some of these limitations for use as cladding in commercial nuclear fuel rods.

During the lifetime of a nuclear fuel assembly, the zirconium alloy fuel rod cladding currently used slowly corrodes, reacting with coolant water to form porous zirconia on the outside of the rod and liberate hydrogen. Much like the iron oxides in everyday experience, the zirconium oxide formed has very little structural strength. During transients, or potential accidents such as Loss of Coolant (LOCA), the rate of corrosion may increase substantially, with an equal increase in the rate of liberation of hydrogen and decrease of load bearing (zircaloy) material. For design purposes, the allowable recession of zircaloy cladding over the lifetime of a fuel assembly, including non-design-basis upsets, is an engineering constraint, as enough material must always remain to carry imposed loads. Current regulations for licensing limit recession to 17% over lifetime, calculated as the total change from beginning of life in cross-sectional area of non-corroded material. Particularly during an event, such as a LOCA, when oxidation is accelerated, the survivability of the fuel rod is intimately linked with the recession.

While SiC has excellent high temperature material properties and has been shown to be stable under neutron irradiation, in order to be considered for deployment in the reactor fleet, the behavior of this material under the limiting conditions known for zirconium alloys must be demonstrated. As well, to provide an incentive this material should exceed the capabilities of zicaloy.

111

The oxidation of SiC and the volatilization of SiO_2 has been investigated for the conditions of the combustion environment by several studies.[2,3,4,5,6] While these studies have included temperature ranges that coincide with the temperatures expected in a commercial light water reactor(LWR) LOCA event, the conditions previously investigated have generally been under low partial pressures of water vapor and low flow rates, contrasting with the conditions existing in a LOCA event. Additionally, due to the limitations of the currently used zirconium alloys, the design-basis LOCA event is limited to short duration (approximately 10 minutes). The previous studies have focused on the long term (many hundreds to thousands of hours) oxidation behavior of SiC at these elevated temperatures.

To prove the viability of advanced ceramic composites as an alternative in LWR's to zircaloy cladding and aid in future licensing efforts, a direct comparison of these two technologies is desired. In contrast to many existing studies of silicon carbide, in a commercial nuclear reactor, the SiC clad would see long-term exposure to relatively low temperatures and short-term exposures to high temperatures. While Barringer[7] and others have begun to investigate SiC corrosion at operating conditions in coolant, some gap in the data still exists for the transient design-basis events, particularly in steam. In this study we investigate this regime and demonstrate the increased margin of safety.

EXPERIMENTAL PROCEDURE

Behavior of Silicon Carbide at LOCA Conditions

Sintered α-SiC (Hexaloy™ SA; St. Gobain Ceramics, Niagara Falls, New York) and CVD β-SiC (TREX Enterprises, Honolulu, Hawaii) test specimens (nominal 22.35mm x 12.07mm x 9.53mm) were used in this investigation.

The primary test stand was a 1700°C tube furnace with a 60cm uniform zone. A 99.8% purity Al_2O_3 6.4 cm ID tube was used as the working section. The SiC test specimens were placed on the upper surface of a 99.8% Al_2O_3 dee tube used as a boat. The specimen boat was inserted into the tube furnace, centered in the tube furnace hot zone. The dee tube was dimensionally matched to the main furnace tube so that the test specimens would be elevated to mid-height of the main furnace tube and subjected to full flow conditions. A platinum-rhodium thermocouple was mounted inside the boat such that the junction was freely exposed to the flow.

Distilled water was added to an aluminum boiler. This container was sealed and the distilled water was brought to a boil, with the resulting pressure driving the system. The vapor was then transported through a stainless steel tube to a sealed stainless steel plenum section immediately upstream of and connected to main test section alumina tube. Between the boiler section and the plenum, the water vapor saturation level was decreased by passing the stainless steel tube through the heated section of a small furnace held at 600°C. At the plenum the water vapor was further dried to allow superheating and to prevent any liquid condensate from forming on the upstream portions of the alumina tube test section away from and upstream of the heated zone of the test furnace. From the plenum, the vapor entered the alumina tube of the horizontal tube furnace, superheating to test conditions and flowing through to the heated test section, finally exhausting to atmospheric conditions. A nickel-chromium thermocouple was used to monitor the temperature of the water vapor entering the tube from the plenum to ensure a temperature of 115°C or above. Two thermocouples were used to monitor the temperature of the alumina tube and its thermal gradient upstream of the test section. Water vapor flow was confirmed visually by observing refractive changes at the exhaust. Water vapor flow rate was confirmed and monitored by the use of the platinum-rhodium thermocouple mounted in the lower part of the specimen boat. Water vapor flow rate was maintained by keeping a constant difference between the oven temperature control thermocouple (not exposed to flow) and the flow monitoring thermocouple, and adjusted by increasing or decreasing the steam back pressure at the boiler.

The distilled water amount was weighed before and after testing to an accuracy of 0.1 lb. Condensate was drained from the plenum, collected, and measured to an accuracy of 10mL The mass of each sample was measured within 0.1 mg before and after exposure testing. Also, before and after each test, twelve measurements of the physical dimensions of each block were taken within 0.01mm, resulting in fine measurement of the surface area of each face of the hexahedron.

The samples were placed into the tube furnace open to atmosphere at one end. The oven was ramped to the test temperature at a rate of 150°C/hr, with a 20 minute soak at test temperature for tests with flow. One set of specimens were tested in non-flowing conditions in air at 1200°C for 6 minutes and one set for 8 hours. One set were tested at 1600°C for 8 hours. Specimens were tested in flowing steam at 1200°C for 10 minutes, 2 hours, 4 hours, and 6 hours. Specimens were testing in flowing steam at 1400°C for 4 hours and 8 hours. The cooldown from test temperatures was held constant at 150°C/hr.

Silicon Carbide Behavior Data Reduction, Oxide Layer Growth

Once the post-test mass was determined, the conversion of SiC (molecular weight 40) to SiO_2 (molecular weight 60) under non-flow conditions was calculated from the weight gain of the samples. Further, by use of the theoretical density of SiC (3.1 g/cm^3), the total volume of SiC reacted was calculated. The face of the specimen resting on the Al_2O_3 boat was assumed not be reacted. From the measurement of the face surface areas prior to test and assuming an equal distribution over all exposed surface areas, the depth of the oxidation front relative to the pre-test material was calculated. The thickness of the oxide layer was also calculated by use of the density of SiO_2 (2.3 g/cm^3) and the exposed surface area.

Silicon Carbide Behavior Data Reduction, Calculation of Recession

It was assumed that during the flow of water vapor, the volatilization of SiO_2 to $Si(OH)_4$ was rapid enough to dominate over the rate of oxidation of SiC to SiO_2. Furthermore, it was assumed that the rate was rapid enough to additionally etch any oxidation occurring prior to the initiation of water vapor flow. All SiO_2 formation on a specimen subjected to steam flow was therefore assumed to occur after the flow of water vapor ceased, during the cooldown of the system in atmospheric air.

For a given test temperature, the amount of SiO_2 formed after cessation of flow was assumed equal to the amount of SiO_2 formed in the 8 hour exposure test in stagnant air at the same test temperature. This assumption results in an oxidation depth one order of magnitude higher when compared to the rates from Ervin.[8] The estimate used in this report, based on a maximum possible upper bound, however, maximizes the amount of SiC reacted for a given test, and thus maximizes the estimate of the amount of hydrogen produced and material recession. Consistent with the definition used in the nuclear power industry, material recession is calculated by the total cross-sectional area change of the test sample.

The total mass loss measured from the experiment was adjusted to a mass loss due to hydrolysis by the subtracting the mass assumed gained during oxidation. Using the density of SiC, a total volumetric change due to hydrolysis was calculated, and from the exposed surface area as measured prior to testing, the depth of hydrolysis reaction was calculated. The calculation of conversion of SiC to SiO_2 was made as previously described. The total reaction front depth was calculated by adding the depth of the hydrolysis front and depth of the oxidation front.

Silicon Carbide Behavior Data Reduction, Calculation of Hydrogen Liberated

Opilia[2,9] and others have established the primary reactions of silicon carbide in the presence of oxygen and water vapor to be

$$SiC + 3/2\ O_2(g) = SiO_2 + CO(g) \tag{1}$$

$$SiC + 3H2O(g) = SiO_2 + 3H_2(g) + CO(g) \tag{2}$$

while the volatilization equations are

$$SiO_2 + H_2O(g) = SiO(OH)_2(g) \tag{3}$$

$$SiO_2 + 2H_2O(g) = Si(OH)_4(g) \tag{4}$$

$$SiO_2 + 3H_2O(g) = Si_2O(OH)_6(g) \tag{5}$$

Equation (2) is used for oxidation calculations in this study, which assumes that no free oxygen (or hydrogen) exists in the stream. From this equation, and considering H_2 liberated as an ideal gas, the volume amount of H_2, expressed as mL at standard temperature and pressure, can be calculated. Examining the volatilization equations, no hydrogen is produced, so only the amount of SiC converted to SiO_2 is of concern.

Zirconium Alloy Calculations of Recession and Hydrogen Liberated

Zirconium alloys have been widely studied and well characterized in the chemistry and conditions of light water reactors. The data of Lemmon[10] and Baker[11] are used here to calculate the recession of zircaloy-2 and the resulting hydrogen liberation.

RESULTS AND DISCUSSION

The percentage weight change for samples tested in air at 1200°C and 1600°C is shown in Figure 1. For the relatively short time durations of this test as compared to other work on corrosion, the change for sintered α-SiC and CVD β-SiC are found to be identical within the error of measurement.

The lower data point at 6 minutes in Figure 1 is considered outlying. No observed phenomenon accounted for this change. It is speculated that this represents some contaminant not carefully cleaned from the surface that was volatilized during the heatup. Ervin[8] showed a significant correlation, as one might expect, in the percentage weight gain with the ratio of surface area to mass. When considering the ratio in these samples, the values measured are comparable to what is expected. Though a linear fit, or even a fit of the parabolic constant, can be made over this range for a constant temperature, the difference between this fit and a constant weight gain based on the maximum gain at the maximum time was within the data spread of the samples.

Figure 2 shows the raw percentage weight change for samples exposed to LOCA steam conditions and a linear curve fits for α-SiC at both 1200°C and 1400°C. Interesting to note are again the outlying data points at 1200°C at a shorter time frame (120 minutes). These, again, have no known cause. The data at 1400°C on CVD β-SiC shows a fairly considerable spread. Both of these samples were tested at the same time and show the trend that was observed in all cases when multiple samples were tested, that the downstream item showed a lower weight loss than its immediate upstream predecessor. The average of the two data points is believable and suggests a substantial decrease in oxidation of CVD β-SiC.

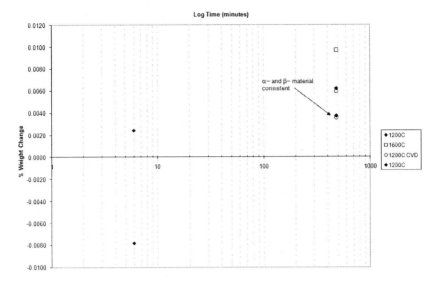

Figure 1: Percent weight change in silicon carbide samples in air.

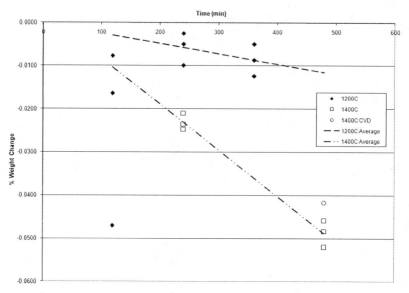

Figure 2: Percentage weight change for SiC specimens tested in superheated steam.

The mass changes shown in Figure 2 seem to be of the proper order of magnitude when compared with the data reported by Opila[4] for 95% pH$_2$O, though when compared to the parabolic rate constants reported there, the value of the present study is several times greater. This cannot be attributed to the higher flow rates (~1 m/s) achieved in this test compared to the 3 cm/min rates maintained by Opila. Tortorelli asserts that recession rates are dependent only upon temperature and water vapor pressure.[6]

The values in Table I were derived from the raw data shown in Figure 1 and Figure 2.

Table I: Experimentally Obtained Values of SiC Oxidation and Volatilization

Temp (°C)	Time (min)	Δw_{oxi} (mg/cm^2)	Δw_{hydr} (mg/cm^2)	depth of reaction, x (μm)
1200	240	0.045	0.102	0.620
1200	360	0.045	0.131	0.712
1400	240	0.072	0.301	1.431
1400	480	0.072	0.554	2.249

Figure 3, Figure 4, and Table II relate the calculated values of Recession and Hydrogen evolution for typical 17x17 fuel rod clad geometry. The geometry used for zircaloy is a standard 17x17 fuel clad, 9.5cm OD x 8.2cm thickness. The geometry used for SiC is the geometry achieved in the CTP Round 7 TRIPLEX specimens. The recession and hydrogen generation in zircaloy are calculated to be over two orders of magnitude higher than that for SiC. This is a striking result, with the implications of the potential for a exceptional gain in safety margin for LOCA conditions.

Table II: Summary of Recession and Hydrogen Liberation Calculated for Zircaloy Tube and Silicon Carbide Tube under LOCA Conditions

Temp (°C)	Time (min)	Recession, SiC CTP Round7 Dimensions	Recession, Zircaloy-2 17x17	H2 Liberation, SiC - CTP Round7 Dimensions (mL / cm^2)	H2 Liberation, Zircaloy-2 17x17 (mL / cm^2)
1200	240	0.032%	28.2%	0.323	134.16
1200	360	0.037%	34.4%	0.371	164.32
1400	240	0.074%	42.1%	0.746	201.99
1400	480	0.116%	58.9%	1.172	285.66

Opila[3] showed that the presence of a high purity alumina reaction tube accelerates the oxidation of SiC in the presence of water vapor. Based on the rate constants determined in that work, the recession of SiC at 1200°C is 1.5 times greater under the influence of an Al_2O_3 reaction tube, while at 1400°C, this factor is 3.4. This may explain some of the higher rate of oxidation previously discussed in Figure 2. Though the tests reported here were conducted under the influence of an Al_2O_3 tube, the SiC recessions derived are not adjusted by these factors, and thus represent a greater calculated oxidation than should be observed in the intended application. For the purposes of engineering comparison to existing zircaloy technologies, this represents a conservative factor.

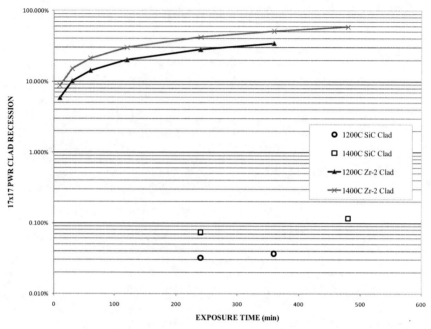

Figure 3: Projected recession rates in 17x17 fuel rod clad. The upper curves represent the clad recession expected based on well-characterized zircaloy-2 at 1200°C and 1400°C for various exposure times. The lower data points represent calculated values for SiC based on the test data of this report adjusted to the TRIPLEX clad dimensions as currently developed by Ceramic Tubular Products.

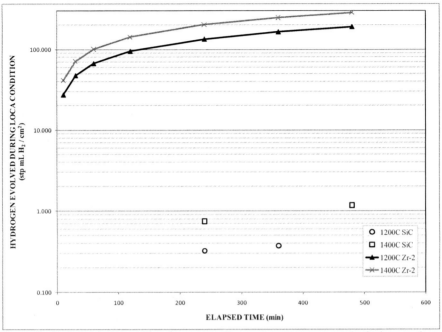

Figure 4: Projected hydrogen liberation rates. The upper curves represent the hydrogen release expected based on well-characterized zircaloy-2 at 1200°C and 1400°C for various exposure times. The lower data points represent calculated values for SiC based on the test data of this report.

SUMMARY

Both high purity α- and β- silicon carbide showed very low corrosion rates when exposed to steam at the temperature of a design basis Loss of Coolant Accident. The experiment readily demonstrated that SiC cladding can withstand steam at the currently allowed design basis LOCA temperatures (1200°C) for periods in excess of 8 hours without any serious loss of material, compared to the limiting case of approximately 10 minutes for zircaloy. The experiment showed that silicon carbide clad can also withstand 1400°C steam in excess of 8 hours. Applying the rates of reaction determined in the present study to the dimensions of the TRIPLEX clad under development for pressurized water reactor 17x17 fuel and comparing this to the known rates of zircaloy, the recession rate was determined to be 800 times less at 1200°C and 500 times less at 1400°C, substantial margin over current allowables. For the evolution of hydrogen, SiC exibited roughly 400 times less per unit surface area than zircaloy at 1200°C and 100 times less at 1400°C. For such cases as the recent Fukushima accident, such margins would have been desirable.

ACKNOWLEDGEMENTS

The authors would like to extend their gratitude to Suzanne Zeismann, PhD for her assistance and insight; Cheryl Lindeman, PhD for her microscopy work and training; Peng Xu of Westinghouse Electric Company, LLC for his suggestions; and the Central Virginia Governor's School for Science and Technology for access to their scanning electron microscope. This work was sponsored by the US Department of Energy under Small Business Innovative Research Grant DE-SC0004225.

REFERENCES
1. K. E. Yeuh, D. C. Carpenter, H. Feinroth, "Clad in Clay," Nuclear Engineering International March 2010.
2. E. J. Opila, R. E. Hann, "Paralinear Oxidation of CVD SiC in Water Vapor," J. Am. Ceram. Soc. 80 [1] 197-205 (1997).
3. E. J. Opila, "Oxidation Kinetics of Chemically Vapor Deposited Silicon Carbide in Wet Oxygen," J. Am. Ceram. Soc. 77 [3] 730-736 (1994).
4. E. J. Opila, "Variation of the Oxidation Rate of Silicon Carbide with Water-Vapor Pressure," J. Am. Ceram. Soc. 82 [3] 625-636 (1999).
5. K. L. More, P. F. Tortorelli, M. K. Ferber, J. R. Keiser, "Observations of Accelerated Silicon Carbide Recession by Oxidation at High Water-Vapor Pressures," J. Am. Ceram. Soc. 83 [1] 211-213 (2000).
6. P. F. Tortorelli, K. L. More, "Effects of High Water-Vapor Pressure on Oxidation of Silicon Carbide at 1200°C," J. Am. Ceram. Soc. 86 [8] 1249-1255 (2003).
7. E. Barringer, Z. Faiztompkins, H. Feinroth, "Corrosion of CVD Silicon Carbide in 500°C Supercritical Water," J. Am. Ceram. Soc. 90 [1] 315-318 (2007).
8. G. Ervin, Jr., "Oxidation Behavior of Silicon Carbide," J. Am. Ceram. Soc. 41 [9] 347-352 (1958))
9. N. S . Jacobson, D. S. Fox, E. J. Opila "High temperature oxidation of ceramic matrix composites," Pure & Appl. Chem. 70 [2] 493-500 (1998).
10. A. W. Lemmon, "Studies Relating to the Reaction Between Zirconium and Water at High Temperatures," Battelle Memorial Institute Report No. BMI-1154 (1957).
11. L. Baker, L. Just, "Studies of Metal-Water Reactions at High Temperatures III. Experimental and Theoretical Studies Of The Zirconium-Water Reaction," Argonne National Laboratory ANL-6548 (1962).
12. K. L. More, P. F. Tortorelli, M. K. Ferber, J. R. Keiser, "Observations of Accelerated Silicon Carbide Recession by Oxidation at High Water-Vapor Pressures," J. Am. Ceram. Soc. 83 [1] 211-213 (2000).

THE EFFECT OF TEMPERATURE AND UNIAXIAL PRESSURE ON THE DENSIFICATION BEHAVIOR OF SILICA AEROGEL GRANULES

J. Matyáš*, M. J. Robinson, and G. E. Fryxell
Pacific Northwest National Laboratory
Richland, WA, USA

ABSTRACT

Materials are being developed in the U.S. for the removal and immobilization of radioiodine from the gas streams in a nuclear fuel reprocessing. Silver-functionalized silica aerogel is being investigated because of its high selectivity and sorption capacity for iodine and its possible conversion to a durable silica-based waste form. In the present study, we investigated the effect of pressureless isothermal sintering at temperatures of 900-1400 °C for 2.5-90 min or isothermal hot-pressing at 1200 °C for 2.5 min on densification of raw and silver-functionalized silica aerogel granules with nitrogen sorption and helium pycnometry. Rapid sintering was observed at 1050 and 1200 °C. Pressureless sintering for only 15 min at 1200 °C resulted in almost complete densification: the macropores disappeared, the specific surface area decreased from 1114 m^2/g to 25 m^2/g, pore volume from 7.41×10^{-3} m^3/kg to 9×10^{-5} m^3/kg, and adsorption pore size from 18.7 to 7 nm. The skeletal density of sintered granules was similar to the bulk density of amorphous silica (2.2×10^3 kg/m^3). Low-pressure hot-pressing accelerated the sintering process, decreasing significantly the pore size and pore volume.

INTRODUCTION

Reprocessing of used nuclear fuel is being considered in the US.[1] In reprocessing, valuable materials are recovered for recycle and the volume of highly radioactive material for storage or disposal is reduced. However, volatile radionuclides, such as tritium (3H), carbon-14 (^{14}C), krypton-85 (^{85}Kr), and iodine-129 (^{129}I), are released into off-gas stream in the plant and must be captured to meet regulatory limits. Of these radionuclides, ^{129}I poses the greatest health risk as it has the half-life of 1.6×10^7 years and is highly mobile in the environment. In the US, silver-exchanged (AgZ) or silver-reduced (Ag^0Z) mordenites are being considered for the removal of ^{129}I from the gas streams. To immobilize iodine, the Ag-bearing mordenites would be mixed with cement and packaged for disposal.

The U.S. Department of Energy is currently investigating alternative sorbents for removing iodine. These sorbents have higher affinity and sorption capacity for iodine and can be directly converted to a final waste form without other materials added. One of these new sorbents, silver-functionalized silica aerogels, shows great promise as a potential replacement for Ag-bearing mordenites. Previously[2], we showed a maximum sorption capacity for I_2 of 44 mass% and decontamination factors in excess of 310 for an Ag^0-functionalized aerogel. After optimization of the preparation procedure, these numbers were increased recently to 48 mass% and 10 000, respectively.[3] We also showed that the iodine-loaded Ag^0-functionalized aerogel could be rapidly consolidated under low pressure and at 1200 °C without significant loss of I_2. In these preliminary experiments, more than 92% of the I_2 was retained in the densified silica-based waste form.[2] We believe that iodine retention can be significantly increased by optimizing of the sintering process.

There is ample evidence in the literature that silica glass can be produced by densifying silica aerogels or xerogels at temperatures well below the melting point of silica. Grandi et al.[4] studied the densification of silica aerogel and xerogels monoliths at temperatures up to 1200 °C and concluded that aerogel led to crack-free glass while xerogels tend to crack and bloat during thermal treatments. Dieudonné et al.[5] investigated the impact of room temperature isostatic compression or thermal sintering at temperatures between 920 and 1100 °C on the porous texture of silica aerogel. The compression at room temperature resulted in a decrease in pore size, but, unlike the thermal sintering, the surface area and the mean size of the fractal clusters that constitute the aerogels remained constant. Perin et al.[6] showed that isostatic compression of monolithic silica aerogel samples with a mercury porosimeter followed by heat-treatment at 1050 °C considerably accelerated the sintering process. Folgar et al.[7] evaluated the changes in microstructure for silica aerogels monoliths at the temperature range between room temperature and 1500 °C. Aerogels heated at 5 °C/min from room temperature to 1100 °C reached the density of amorphous silica (2.2×10^3 kg/m^3) and exhibited a pore volume of less than 0.1×10^{-3} m^3/kg. Scherer et al.[8] used nitrogen sorption to follow the structural evolution of small pieces of silica aerogel sintered at 1050 °C for periods ranging from 0.5 to 39 h. The resulting pore size distribution was incorporated into the cylinder model[9] that provided an accurate prediction of densification kinetics, including the changes in surface area and pore diameter.

The present study was conducted to investigate densification of raw and Ag^0-functionalized silica aerogel granules during pressureless and uniaxial hot-press isothermal sintering, and identify optimal conditions for consolidating this highly porous material into a durable silica glass. The thermally sintered samples were analyzed with Brunauer, Emmett, Teller (BET) gas adsorption analysis for surface area, pore volume and pore size distribution (BJH); with helium pycnometry for skeletal density (the ratio of the sample mass to skeletal volume, which consists of the volumes of the solid material and closed pores); with scanning electron microscopy for particle morphologies and elemental distribution; and with stereomicroscope for surface morphology of granules.

MATERIALS

The hydrophobic raw silica aerogel granules were purchased from United Nuclear (Laingburgh, MI). They are hydrophobic because of the presence of thrimethylsilyl groups on the porous surface. The granules (1-4 mm) were used as received (raw aerogel) or heat-treated at 5 °C/min from 20 to 400 °C with 1 h hold at 400 °C (400C aerogel) or chemically modified with silver nanoparticles with the procedure developed earlier[1] (Ag^0-functionalized aerogel). Briefly, the silver nanoparticles were produced on the propylthiol-modified pore surfaces by reducing the silver thiolate adduct ions at 165 °C for 2 h under flowing 2.7% H$_2$ in Ar.

EXPERIMENTAL AND ANALYTICAL METHODS
Pressureless thermal sintering

Small samples of raw and Ag^0-functionalized silica aerogel granules were heat-treated in 1-mL Pt-crucibles and in a high-temperature furnace (Deltech Inc, Denver, CO) at temperatures ranging from 900 °C to 1400 °C for various times from 2.5 to 90 min.

Uniaxial hot pressing

Figure 1 shows a schematic of the uniaxial hot press system that was designed and fabricated in our laboratory. In this experimental method, the pressure and heat were applied simultaneously to enhance the densification rate of the granulitic silica aerogel samples inside an alumina crucible. First, the aerogel granules were loaded into alumina crucible (9.55 mm OD × 6.42 mm ID × 19.10 mm tall) with the help of ceramic syringe, which is composed of an alumina tube of the same inner diameter as the sample crucible and 0.3 mm thick alumina disk that sits on a short alumina plunger rod (Ø 6.37 mm). The granules were subsequently compressed with 6-kg weight on the plunger rod. The crucible with compressed granules was secured with a Pt wire to the alumina push rod, which was mounted to the Duramaster Rod Cylinder (Greenco Mfg. Corporation, Tampa, FL) outside of the furnace. This Rod Cylinder was pressurized with air to the desired force on the sample (see Table 1), extending a push rod and compressing the sample in the crucible against mullite block inside of the furnace that was pre-heated to 1200 °C. The test duration started when contact was made between the crucible and the mullite stage. After a defined time period, the rod was retracted from the furnace, the crucible was removed and sample air-quenched on an alumina plate.

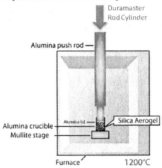

Figure 1. A schematic of the uniaxial hot press system.

The force exerted by a Duramaster Rod Cylinder can be expressed as $F = pA_b = p\pi d_b^2/4$, where F is an exerted force (N), p is an input air pressure (Pa), A_b is an effective bore area ($A_b = 3.167\times10^{-3}$ m^2) and d_b is an effective bore piston diameter ($d_b = 6.35\times10^{-2}$ m). The push rod applies this force to compress the sample in the crucible. The pressure on the sample is defined as $P_s = F/S_s$, where S_s is the effective area of the sample crucible ($S_s = 3.237\times10^{-5}$ m^2). The pressure exerted by the sample crucible on the mullite base is defined as $P_c = F/S_c$, where S_c is the area of the bottom of the crucible ($S_c = 7.163\times10^{-5}$ m^2). Table 1 shows the calculated values for the forces generated by a pneumatic ram, the pressures exerted on the sample and the mullite base as a function of the input air pressure.

Table 1. Forces generated by a pneumatic ram (F) and the pressures exerted on the sample (P_s) and the mullite base (P_c) as a function of the input air pressure.

Input air pressure, MPa	F, N	P_s, MPa	P_c, MPa
0.12	382	11.8	5.3
0.29	928	28.7	13.0

Simultaneous thermogravimetry and mass spectrometry

A Netzsch STA 409 C/CD (Netzsch, Boston, MA) coupled to a skimmer mass spectrometer system QMS 403/5 (Netzsch, Boston, MA) was used to determine the mass loss and identity of the species being evolved during heating of silica aerogel granules. Samples of raw and heat-treated silica aerogels (~ 20 mg each; 400C aerogel; 20 to 400 °C at 5 °C/min, 400 °C for 1 h) and Ag^0-functionalized silica aerogel (14 mg) were heated under flowing He (20 mL/min) at 5 °C/min from 25 to 100 °C, held for 2 h, and then at 5 °C/min to 1200 °C.

BET gas adsorption analysis

The analysis was performed on samples degassed at 25 °C under vacuum. The data from nitrogen adsorption/desorption at 77 K were collected with a Quantachrome Autosorb 6-B gas sorption system (Quantachrome Instruments, Boynton Beach, FL). The surface area was determined from the isotherm with the five-point BET method. The Barrett–Joyner–Halenda (BJH) method was used to calculate the total pore volume and pore size distribution including average pore diameter. The accuracy of BET analysis was about 5% as determined by multiple analyses of the same samples.

Helium pycnometry

A helium pycnometer (AccuPyc II 1340, Micromeritics Inc., Norcross, GA) was used to determine the skeletal volume (the volume of the solid material excluding surface-connected porosity, but including closed pores) of the silica aerogel samples by measuring the pressure change of helium (99.995% pure) in a calibrated volume. The skeletal density of samples was then calculated from the known sample masses. The pycnometer was operated with 1 mL sample cell module with an additional insert which reduced the volume to 0.14 mL. The volume of the insert was calibrated with a 0.1023 ± 0.0001 mL tungsten cylinder. The calibration was checked with tungsten and aluminum samples of known density[10], 19.3×10^3 kg/m^3 and 2.7×10^3 kg/m^3 at 20 °C, respectively. The measured densities $19.3501 \pm 0.0248 \times 10^3$ kg/m^3 and $2.7023 \pm 0.0152 \times 10^3$ kg/m^3 were in good agreement with postulated values.

The samples were weighed on a Sartorius analytical balance (Brinkmann Instruments Co., Westbury, New York) to the nearest 0.1 mg. Sample masses varied between 8 to 60 mg depending on the aerogel apparent density. The insert with samples was purged (cell filling/expulsion of helium) 300 times to clean up the sample and remove air and moisture from inside the chamber. This was followed by 10 measurement cycles of the sample volume for density determination, the results of which were averaged.

Scanning electron microscopy

Samples of hot pressed Ag^0-functionalized silica aerogel were analyzed with thermal field emission scanning electron microscope (FESEM, JEOL 7001F, JEOL, Boston, MA) equipped with an energy dispersive silicon drift detector (APOLLO XL, EDAX, Mahwah, NJ) with 30 mm^2 active area and 35° takeoff angle. Samples were mounted in epoxy, cross-sectioned, polished, and placed in the microscope without carbon coating. Images and elemental analyses were collected in the low vacuum mode (2.7 Pa), at a 10-mm working distance, with a 10 kV accelerating voltage, and with a medium probe size 15. Images were collected with a back scatter electron detector at the nose of the pole piece. A 512×400 pixel (pixel size 1.087 μm) elemental spectrum dot map was collected at a magnification of 230X, with a dwell time of 200 μsec/pixel. Each map was a compilation of 100 scans with drift correction. Each elemental map was then processed to scale brightness to atomic %.

RESULTS AND DISCUSSION

Figure 2 shows the mass loss and detected ion current for methyl groups (from the thrimethylsilyl groups) as a function of the temperature for raw and heat-treated silica aerogels. The raw aerogel lost 0.27 mass% of physically sorbed water during the heating from room temperature to 100 °C. An additional water loss of 1.54 mass% was observed during the 2 h long hold at 100 °C. The loss of 13 mass% between 250 and 900 °C is attributed to the removal of hydroxyl and thrimethylsilyl groups from aerogel surface. The continuous mass loss at temperatures above 900 °C corresponds to the removal of the remaining hydroxyl groups. The heat-treated aerogel contained only 0.14 mass% of physisorbed water and exhibited a mass loss of 4.9% between 250 °C and 900 °C. By comparing the area under the peaks attributed to the methyl groups for raw and heat-treated aerogel approximately 61% of the methyl groups were removed during the ramp-heating and 1-h hold at 400 °C.

Figure 3 shows the mass loss and detected ion current for methyl and thiol groups (from the propylthiol monolayer on the aerogel surface prior to functionalization with silver) as a function of temperature for Ag⁰-functionalized silica aerogel. This aerogel contained water at 0.3 mass%. The major mass loss of 14.2 mass% from 210 to 840 °C is attributed to the decomposition of propylthiol monolayer. An additional decrease in mass at higher temperatures corresponded to the removal of hydroxyl groups.

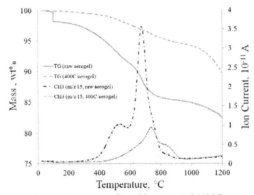

Figure 2. TG-MS curves for raw (raw aerogel) and heat-treated (400C aerogel) silica aerogels as a function of the temperature.

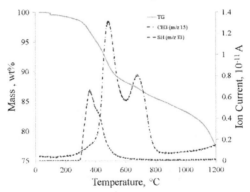

Figure 3. TG-MS curves for Ag⁰-functionalized silica aerogel as a function of the temperature.

Figure 4 shows the changes in the surface area, total pore volume, and average pore size for adsorption and desorption for raw aerogel granules and for granules that were pressureless sintered at 900, 1050, and 1200 °C for 2.5 min. Figure 5 illustrates with optical images the morphology of the granules before and after pressureless sintering. The raw granules exhibited the surface area 1114 m^2/g, pore volume 7.41×10^{-3} m^3/kg with adsorption pore sizes 18.7 and >50 nm, and desorption pore size 6.5 nm. The adsorption pore size >50 nm indicated that some macropores are present in the aerogel. The granules sintered at 900 °C showed little sign of densification with the surface area 1051 m^2/g because of the predominant collapse of macropores. These pores collapsed first. The lack of densification is readily seen in image B (Figure 5). No apparent change in the size and shape of granules was observed at this temperature, but granules were mechanically strengthened (observed qualitatively) and converted to rigid porous glass granules. The higher total pore volume, 7.73×10^{-3} m^3/kg, and adsorption and desorption pore sizes 33.5 and 23.9 nm, respectively, compared to raw aerogel can be rationalized by the loss of trimethylsilyl and silanol groups from the pore surfaces and smoothing of pore walls during the sintering.

A significant decrease in specific surface area, pore volume, and size was observed for granules subjected to pressureless sintering at 1050 and 1200 °C. After 2.5 min at 1050 °C the surface area was 724 m^2/g and at 1200 °C 397 m^2/g. The pore volume and average pore size followed the same trend. The pore volume decreased to 3.58×10^{-3} and 1.45×10^{-3} m^3/kg, adsorption pore size to 24 and 18.7 nm and desorption pore size to 18.6 and 12.6 nm. While there was some shrinkage of granules at 1050 °C (image C Figure 5), high shrinkage was observed at 1200 °C (image D Figure 5).

The rapid sintering at 1200 °C was demonstrated with pressureless sintering for longer times. Figure 6 shows the effects of sintering time on densification of raw silica aerogel granules during sintering at 1200 °C for 2.5, 5, and 15 min. Compared to 2.5 min the thermal sintering for 5 min decreased the surface area more than 5-fold and pore volume more than 6-fold to 68 m^2/g and 0.23×10^{-3} m^3/kg, respectively. An additional 10 min at this temperature further decreased the surface area to 25 m^2/g and pore volume to 9×10^{-5} m^3/kg and uniform adsorption and desorption pore sizes 7.5 and 7.4 nm. Figure 7 shows the optical images of the granules before and after pressureless sintering at 1200 °C. With prolonged sintering, larger glass particles formed as individual sintered granules agglomerated after 2.5 min. These particles contained a small amount of microporosity and were opaque because of the small scale foaming from trapped hydroxyl groups.

Figure 8 illustrates the effect of temperature and time during pressureless sintering on skeletal density of silica aerogel granules. The difference in density observed between raw silica aerogel granules, $1.6509 \pm 0.0095 \times 10^3$ kg/m^3, and amorphous silica[11], 2.2×10^3 kg/m^3, can be explained by the steric effect and lower bulk density of the trimethylsilyl groups on the surface. This effect was almost eliminated and the accessibility of pores by He was improved by the partial removal of these groups during 2.5-min sintering at 900 °C, resulting in a density of $1.9033 \pm 0.0603 \times 10^3$ kg/m^3. Sample density increased to $2.1391 \pm 0.0568 \times 10^3$ kg/m^3 with an additional 2.5 min of sintering, indicating complete removal of organic groups from the pore surfaces. Even longer sintering times produced granules with densities around 2.2×10^3 kg/m^3. At higher temperatures except for 1400 °C, the densities close to amorphous silica were achieved after 2.5 min. Small scale bloating of the granules occurred at 1200 °C and was significantly more pronounced at 1300 and 1400 °C. This led to lower densities. The bloating at 1400 °C observed during the first 15 min, which was responsible for the densities lower than that of the raw aerogel, was completely eliminated after 40 min of sintering. In contrast, at 1300 °C the bloating persisted even after 90 min and resulted in density $1.7177 \pm 0.0155 \times 10^3$ kg/m^3. The excessive bloating at this temperature is illustrated in Figure 9.

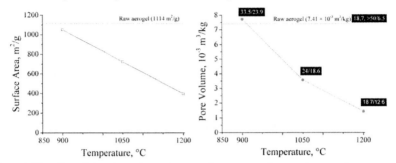

Figure 4. The effect of sintering temperature on surface area, total pore volume, and average pore size for adsorption and desorption (black box; Ads/Des, nm) for raw silica aerogel granules after 2.5 min pressureless sintering at 900, 1050, and 1200 °C.

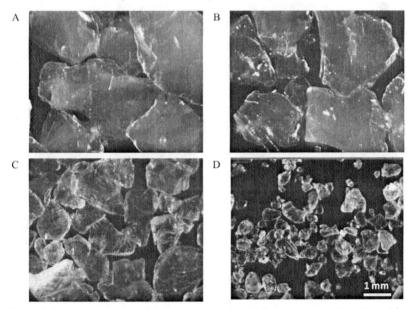

Figure 5. Optical images of raw silica aerogel granules (A) that were pressureless-sintered for 2.5 min at 900 °C (B), 1050 °C (C), and 1200 °C (D).

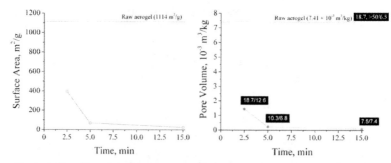

Figure 6. The effect of sintering time on surface area, total pore volume, and average pore size for adsorption and desorption for raw silica aerogel granules after pressureless sintering at 1200 °C.

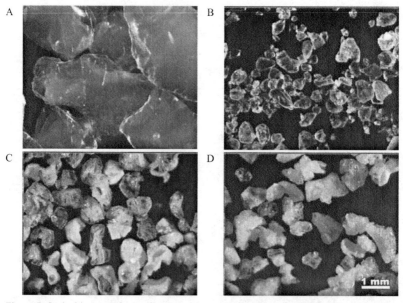

Figure 7. Optical images of raw silica aerogel granules (A) that were pressureless-sintered at 1200 °C for 2.5 min (B), 5 min (C), and 15 min (D).

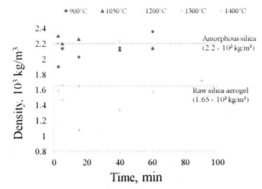

Figure 8. The effect of sintering temperature and time on skeletal density of silica aerogel granules.

Figure 9. Optical image of bloated silica aerogel granule after heat-treatment at 1300 °C for 15 min.

Figure 10 A shows the adsorption pore size distributions for compressed and uncompressed granules of raw silica aerogel. The heat-treatment of untreated aerogel at 1200 °C for 2.5 min resulted in the removal of macropores and decrease of the total pore volume from 7.41×10^{-3} to 1.45×10^{-3} m³/kg. While the maximum pore size decreased to 28.7 nm, the largest fraction of the pores remained at 18.7 nm. Furthermore, the surface area decreased from 1114 to 410 m²/g. The application of 11.8 MPa pressure at the same temperature and time decreased the pore volume to 1.25×10^{-3} m³/kg, broadened the pore size distribution, and increased the surface area to 920 m²/g. The pore size distribution was shifted to smaller pore sizes with most of the pores around 6 nm. In addition, the maximum pore size decreased to 14.3 nm. The observed increase in surface area can be explained by the increase of the surface roughness of nanometer-sized clusters inside the pores. The sample had a skeletal density, $2.2497 \pm 0.022 \times 10^3$ kg/m³, similar to amorphous silica, suggesting the absence of the closed pore volume. Increasing the pressure to 28.7 MPa decreased the pore volume to 0.43×10^{-3} m³/kg and surface area to 569 m²/g. The largest fraction of pores was centered around 2.8 nm and the maximum pore size decreased to 7 nm. The skeletal density of $2.2288 \pm 0.0203 \times 10^3$ kg/m^3 was almost identical to the density of a sample produced at lower pressure. Figure 10 B shows the adsorption pore size distributions for compressed and uncompressed granules of Ag^0-functionalized silica aerogels. The noisy BET data resulted from the use of small samples of low surface area. In contrast to raw aerogels, the untreated Ag^0-functionalized silica aerogel exhibited a surface area 196 m^2/g, total pore volume 0.17×10^{-3} m^3/kg and maximum pore size less than 10 nm, if a few detected macropores are not included. Deposition of the silver nanoparticles resulted in > 5-fold decrease in the specific surface area and > 40-fold decrease in pore volume. After 2.5 min at 1200 °C the surface area and pore volume decreased to 40 m^2/g and 0.07×10^{-3} m^3/kg, respectively. The application of pressure at the same temperature and time resulted only in small changes of the surface area, total pore volume, and pore size distribution with most of the pores centered around 10 nm. The sample sintered at lower pressure had a surface area 23 m^2/g, pore volume 0.6×10^{-3} m^3/kg, and skeletal density 3436.7 ± 40.9 kg/m^3. The pressure of 28.7 MPa resulted in the surface area 35 m^2/g, pore volume 0.6×10^{-3} m^3/kg and skeletal density 3357.5 ± 29.3 kg/m^3. With the knowledge of the sample masses and volumes, the density of silver (10500 kg/m^3)^9, and mathematical regression, we were able to estimate the silver content in the hot-pressed samples. The silver contents were 45.5 and 43.5 mass% in the sintered samples, respectively.

The elemental distribution map of the surface for Ag^0-functionalized silica aerogel that was hot-pressed at 1200 °C for 2.5 min under 11.8 MPa is shown in Figure 11. Small silica rich pieces of aerogel are densified impurities that were introduced inadvertently from a previous batch of raw aerogel granules. The silver, sulfur, and carbon that were introduced onto the surface of the highly porous aerogel backbone at different stages in the functionalization process were uniformly distributed across the surface. The dark vein that starts at the left top corner in Figure 11 is sintered material that contains silver particles that are > 400 nm and more distinctly separated.

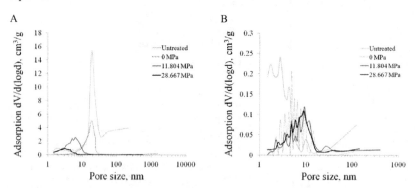

Figure 10. Adsorption log differential pore volume dV/d(logd) vs. pore size for raw aerogel (A) and Ag^0-functionalized aerogel (B); Untreated – room temperature and no pressure, 0 MPa – 1200 °C for 2.5 min and no pressure, 11.8 and 28.7 MPa – hot-pressed at 1200 °C for 2.5 min.

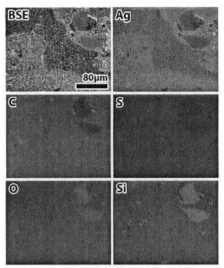

Figure 11. Elemental distribution for hot-pressed Ag^0-functionalized silica aerogel; 1200 °C for 2.5 min under 11.8 MPa.

CONCLUSIONS

There is a possibility that used nuclear fuel could be reprocessed in the US. If that occurs, regulations are in place that limit the release of radionuclides from such a facility. To control the release of ^{129}I in the gas streams from a reprocessing facility, new materials are being investigated to remove iodine. Silver-functionalized silica aerogels have been studied as materials for the removal of iodine and subsequent consolidation to a viable waste form.

The pressureless sintering of raw aerogel granules indicated that the optimal sintering temperature for rapid densification lies in the range 1050-1200 °C. The large-scale bloating of granules caused by entrapped hydroxyl groups and observed at 1300 and 1400 °C was completely eliminated after 40 min at 1400 °C. However, the bloating persisted even after 90 min at 1300 °C resulting in a low skeletal density, 1.7177×10^3 kg/m^3. The short time hot-pressing at 1200 °C allowed the sintering process to be accelerated, decreasing significantly the pore size and volume; the large pores were transformed into numerous small pores.

In the near future, the raw, Ag^0-functionalized, and iodine-loaded aerogel granules will be hot-pressed at temperatures from 1050 to 1200 °C for longer times. In addition, the chemical durability of sintered silica-based materials will be evaluated with product consistency test (PCT).

ACKNOWLEDGEMENTS

Authors would like to thank Xiaohong Shari Li for BET analysis and Jarrod Crum for SEM-EDS investigations. Matthew Robinson is grateful to the DOE Science Undergraduate Laboratory Internship Program for his appointment at PNNL. This work was funded by the U.S.

Department of Energy's Fuel Cycle Research and Development Program. PNNL is operated for DOE by Battelle under Contract DE-AC05-76RL01830.

REFERENCES

[1]*National Nuclear Security Administration, Office of Fissile Materials Disposition*, Report to Congress: Disposition of Surplus Defense Plutonium at Savannah River Site, http://www.nci.org/pdf/doe-pu-2152002.pdf, DOE, Washington DC (2002).

[2]J. Matyáš, G. E. Fryxell, B. J. Busche, K. Wallace, and L. S. Fifield, Functionalized silica aerogels: Advanced materials to capture and immobilize radioactive iodine, in Ceramic Materials for Energy Applications (Eds. H Lin, Y. Katoh, K. M. Fox, I. Belharouak, S. Widjaja, and D. Singh), John Wiley & Sons, Inc., Hoboken, NJ, US, doi: 10.1002/9781118095386.ch3, Ceramic Engineering and Science, **32** (9), 23-33 (2011).

[3]J. Matyáš, Silver-functionalized silica aerogels for capture and immobilization of iodine, Invited presentation at the Fuel Cycle Research and Development Annual Review Meeting at Argonne National Laboratory, Chicago, IL (2011).

[4]S. Grandi, P. Mustarelli, C. Tomasi, G. Sorarù, and G. Spanò, On the relationship between microstructure and densification of silica gels, *Journal of Non-Crystalline Solids*, 343, 71-77 (2004).

[5]Ph. Dieudonné, P. Delord, J. Phalippou, Small angle X-ray scattering of aerogel densification, *Journal of Non-Crystalline Solids*, 225, 220-225 (1998).

[6]L. Perin, S. Calas-Etienne, A. Faivre, and J. Phalippou, Sintering of compressed aerogels, *Journal of Non-Crystalline Solids*, 325, 224-229 (2003).

[7]C. Folgar, D. Folz, C. Suchital, D. Clark, Microstructural evolution in silica aerogel, *Journal of Non-Crystalline Solids*, 353, 1483-1490 (2003).

[8]G. W. Scherer, S. Calas, R. Sempéré, Densification kinetics and structural evolution during sintering of silica aerogel, *Journal of Non-Crystalline Solids*, 240, 118-130, (1998).

[9]G. W. Scherer, Sintering of Low Density Glasses: I. Theory, *Journal of the American Ceramic Society*, **60** (5-6), 236-239 (1977).

[10]R. H. Perry and D. W. Green, Eds., Physical and Chemical Data, Section 3, in "Perry's Chemical Engineers' Handbook," 6th ed., McGraw-Hill, New York (1984).

[11]R. K. Iler, The chemistry of silica: Solubility, polymerization, colloid and surface properties, and biochemistry, Wiley-Interscience, New York (1979).

MICROSTRUCTURAL ANALYSIS OF NUCLEAR GRADE GRAPHITE MATERIALS

Kentaro Takizawa, Toshiaki Fukuda, Akira Kondo, Tokai Carbon Co. LTD., Japan; Yutai Katoh, G E Jellison, Oak Ridge National Laboratory, USA

ABSTRACT

Graphite materials under nuclear services undergo very significant irradiation effects including dimensional instability and modifications in various thermo- physical and mechanical properties[1-4]. Because these changes are resulting from the irradiation-induced structural changes the graphite in nano-, micro- and meso-scopic scales, it is essential to adequately characterize the materials' structures before and after neutron irradiation and to correlate the structural changes with the evolving macroscopic properties. Objective of this work is to characterize the microstructures of the as-fabricate fine-grained nuclear graphite manufactured by Tokai Carbon Co., Ltd.

The graphite consists of filler cokes, binder, pores, and inherent cracks. The characteristic microstructures of the fine-grained isotropic graphite were determined by focusing on the attributes of these elements. Specifically, the sizes, shape factors, and the crystallographic anisotropy were evaluated so that these properties may be correlated with the thermophysical properties, macroscopic anisotropy, fracture properties, and the irradiation effects on them. In future work, these materials will be examined for various properties and microstructural changes by the neutron irradiation in the High Flux Isotope Reactor, Oak Ridge National Laboratory. This paper gives the initial results of microstructural evaluation for the pre-irradiated materials.

INTRODUCTION

The very High Temperature gas-cooled Reactor (VHTR) as one of Generation-IV reactors is an advanced High Temperature Gas-cooled Reactor (HTGR) which is graphite-moderated and helium gas-cooled. It can provide high temperature helium gas about 950 °C to the reactor outlet. It is possible to use this high temperature as heat source not only for power generation but also for hydrogen production.

Graphite components used as a core structure has high heat-resistant and large heat capacity. It is expected to prevent from a rapid temperature change at the accident. From the view point of economy, the graphite components will need their long service life. In that case, it is expected to decrease the reactor core exchange frequency and graphite waste.

Graphite materials under nuclear services cause very significant modifications in various thermal and mechanical properties. These modifications are resulting from the irradiation-induced microstructural changes the graphite[5-8]. It is essential to systematically characterize the microstructures before and after irradiation and to correlate the structural changes with the evolving macroscopic properties. It helps develop the graphite to improved radiation service life.

EXPERIMENTAL

(1)Matrix for Microstructural Analysis

The graphite consists of filler cokes, carbonized binder, pores, and inherent cracks. The characteristic microstructures of graphite are determined by focusing on the attributes of these constituents. Table 1 shows a matrix for microstructural analysis. Each constituent is evaluated by its information of volume, size, shape, orientation and by the degree of graphitization before and after irradiation. Some results of evaluation (red character of Table 1) for as-manufactured graphite are reported in this paper.

Table 1 Matrix for Microstructural Analysis

Constituents	Volume	Size	Shape	Orientation	Degree of Graphitization
Filler	OAM	-	-	OAM^{*3} μXRD	TEM RAMAN
Binder	OAM	-	-	OAM μXRD	TEM RAMAN
Pore	OM^{*1} MP^{*2}	OM MP	OM	-	-
Crack	OM SEM	OM SEM	OM SEM	-	-
Whole	-	-	-	ER CTE	XRD RAMAN

(2)Materials for Evaluation

G347S and G458S manufactured by Tokai Carbon. Co, LTD. were evaluated. These materials are fine-grained isotropic graphites (Fig.1).

Fig.1 Texture of G347S and G458S

Table 2 shows mechanical/thermal properties of G347S and G458S. These materials have different properties reflecting the difference of their texture. G458S has more crystallization parts than G347S. These differences enable to evaluate adequately the relation between microstructure and macroscopic properties.

To evaluate the microstructure, these materials were embedded in an epoxy resin and polished the surface after hardening. Opened pores were filled by the epoxy resin by this method.

Table 2 Mechanical/thermal properties of G347S and G458S

Grade	Bulk density (g/cm^3)	Electrical resistivity (μΩ•m)	Flexural strength (MPa)	Coefficient of thermal expansion (x10^{-6}/°C)	Thermal conductivity (W/m • k)	Young's modulus (GPa)
G347S	1.85	11.0	49.0	4.2* (5.5**)	116	10.8
G458S	1.86	9.5	53.9	3.1* (4.4**)	139	11.3

RESULT AND DISCUSSION

(1)Evaluation of pores

Pore size distribution by Optical Image Analysis

The automated optical image analysis system was used to estimate the pore size and area distribution. Fig.2 shows the composite image of G347S. This composite image consists of 192 images. Each image was obtained at 400 magnifications on the monochromatic 8-bit gray scale, where the brightness of individual pixels ranges between 0 (black) and 255 (white). This 8-bit gray scale image changes into a histogram which illustrates the distribution of shades of gray in a graphical form (Fig.3). As existing pore does not reflect visible light, dark region is recognized as pores in this histogram. The threshold between pores and others is determined experientially and manually like Fig.3 shows. After that, pores are automatically identified (filled in red, Fig.2) and calculated the size and area.

Fig.4 and Fig.5 show the results of pore area distribution by automated optical image analysis. The vertical axis of each graph shows cumulative pore number and cumulative pore area, respectively. Approximately 80% of pores of each material are less than 50μm^2 and the number of pores of G458S is about twice as much as that of G347S. On the other hand, those pores less than 50μm^2 don't influence much of the total pore area. Fig.6 shows the distribution of the aspect ratio of pores. This graph shows that both materials have the similar distribution. Table 3 shows the summary of automated optical image analysis. Each value in this table is the average of 9 measurements. Although G347S has larger pores compared to G458S, the difference between the numbers of pores more than 50μm^2 is quite small. Fig.7 shows the ratio of the total pore area to the surface area for G347S. This graph shows the variation is slightly wide. It is unclear whether the variation of this graph caused by measurement or material itself.

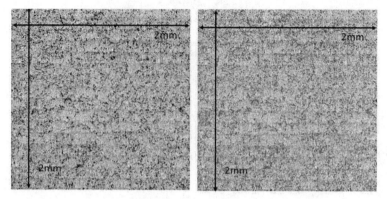

Fig.2 Composite image of G347S

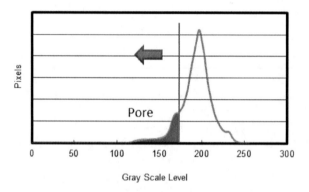

Fig.3 Histogram of 8 bit gray scale image

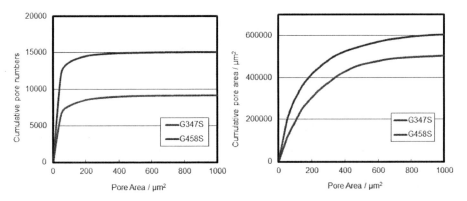

Fig. 4 Cumulative pore numbers   Fig.5 Cumulative pore area

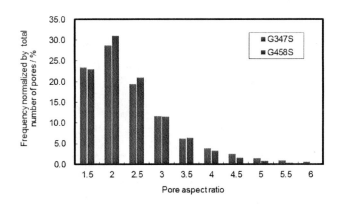

Fig.6 Distribution of pore aspect ratio

Table 3 Summary of optical image analysis

stats	Area / μm^2 / 4.63mm^2					Aspect ratio / -		Feret / μm	
	mean	stdev	Number		Sum	mean	Stdev	mean	stdev
			-	> 50μm^2					
G347S	58.1	2.58	9146	2569	531249.0	2.2	0.02	8.9	0.14
G458S	39.6	1.75	16542	2883	654742.2	2.1	0.01	7.5	0.18

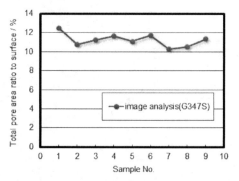

Fig.7 Total pore area ratio to surface

It is necessary to remove the manual influence completely to confirm a cause of variation and obtain the accurate value of the ratio of the total pore area to the surface. By assuming that the distribution of color reflecting each microstructure has normal distribution curve respectively, the histogram obtained in this study can be separated into four normal distribution curves. Each curve attributes to the closed pore, the opened pore, the carbon matrix and the basal plane of carbon matrix, respectively (Fig.8). The total pore area ratio to surface was estimated directly from histogram by using the curve area instead of the value obtained by the automated optical image analysis. Fig.9 shows the result of evaluation using this direct analysis method of G347S. The value obtained by the direct analysis method shows smaller variation and higher value than that by the conventional automated method. G458S also showed the similar tendency. The accuracy of this method is discussed later.

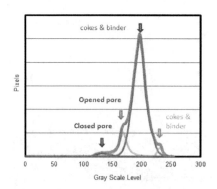

Fig.8 Peak separation of histogram

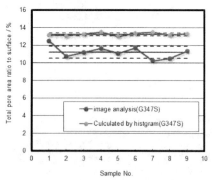

Fig.9 Evaluation using direct analysis
method of G347S

Pore size distribution by Mercury Porosimetry

A mercury porosimetry measurement was used to evaluate the pore size and volume distribution. Fig.10 shows the result of measurements. The value of the pore size obtained by mercury porosimetry is smaller than that obtained by the optical image analysis because the smallest pore diameter can be measured by this method. Both graphite materials have three impregnation peaks. Each peak corresponds to space among particles, structural defects and pores inner particle, respectively. Total pore volume ratio to sample volume is 12.7 % (G347S) and 13.6 % (G458S), respectively. Average pore diameter of G347S is larger than G458S. This tendency is as same as the result obtained by optical image analysis (Table3). Structural defects (cracks) are recognized as pores by optical image analysis. It indicates that the optical image analysis method evaluate pore diameter up to about 0.1μm of the value of mercury porosimetry. Total pore volume ratio to sample volume up to 0.1μm is 10.1 % (G347S) and 10.7 % (G458S), respectively.

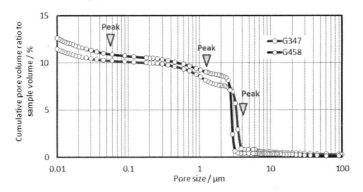

Fig.10 Pore size distribution by mercury porosimetry

Comparison between the two methods using ideal model

Fig.11 shows an ideal model to compare the value obtained by the optical image analysis with the value obtained by the mercury porosimetry. Each cubical unit cell has a same size of a spherical pore corresponded to the actual pore volume ratio. Ideal graphite is considered to be comprised of an array of these cubical unit cells. The information of pore area obtained by the optical image analysis was estimated by cutting this ideal graphite. Fig.12 shows the result of comparison between the two methods using the ideal model. The green line described in Fig.12 is the correlation curve estimated by ideal model. Two open markers indicate the value obtained by the optical image analysis with the conventional method. Other two solid markers indicate the value obtained by the optical image analysis with the direct analysis method. Actual graphite involves a large number of small pores compared to the ideal graphite like Table 3 shows. It indicates that the actual plots are considered to be above the ideal correlation curve. The open marker of G347S is however below the ideal curve. It indicates that the value of total pore area obtained by the optical image analysis with the conventional method is considered to be underestimated. On the other hand, both values obtained by the optical image analysis with the direct analysis method are above the ideal curve. The value of G458S is further above the ideal curve than that of G347S. It corresponds to the result that G458 has more small size of pores than G4347S. These results indicate that the value obtained by the direct analysis method is more reasonable to estimate the total pore area.

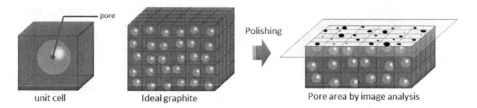

Fig.11 Ideal model of graphite with pore

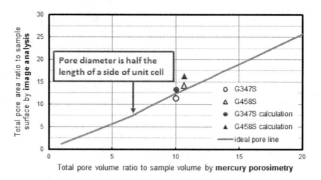

Fig.12 Comparison between two methods using the ideal model

(2)Evaluation of crystal

Derivation of the relations

There have been a number of experimental studies on Raman scattering of micro-crystallite graphite. First-order Raman spectra of micro-crystallite graphite show a disorder-induced band around 1360cm^{-1} (D-band) in addition to the in-plane E_{2g} zone center mode around 1580cm^{-1} (G-band) [9]. D-band is known as the peak that is attributed to symmetry breaking due to the finite crystal size. Therefore the ratio of D-band to G-band intensity is closely related with the finite crystal size [10]. Some relations between the ratio of D-band to G-band intensity and the crystallite size La which is obtained from X-ray diffraction (XRD) have been reported[11-13]. On the other hand, the value of the full width at half maximum of G-band (G-FWHM or $\Delta 1580$) is also known as graphitization parameter.

The same polished sample of as optical image analysis was used. Graphite materials usually grind into powder for measurement on Raman microscopy. However, this resin form is more appropriate in view of post irradiation examination dealing with radioactive graphite.

The polished graphite, in other words edge rich graphite, has quite different spectrum from non-polished graphite because D-band is attributed to symmetry breaking caused by polishing[14]. In this paper, we derive the relations between graphitization parameter obtained by Raman using polished graphite and the crystallite size La.

The Raman spectra were measured by a HORIBA HR-800 spectrometer, using the 532nm line of Nd:YAG laser with a maximum power of 100mW. The 532nm pulsed laser beam was focused to $100\mu m^2$ using vibration mirror to obtain the average spectra of graphite materials.

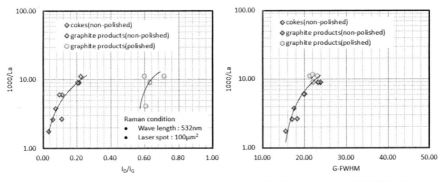

Fig.13 Comparison I_D/I_G with
crystallite size La

Fig.14 Comparison G-FWHM with
crystallite size La

Fig.13 shows the result of comparison the ratio of D-band to G-band intensity with crystallite size La obtained by XRD. Some results using other carbon materials are plotted in the same graph to compare with the polished graphite. Polished graphite has different correlation curve from other materials because of the influence due to polish. It is inappropriate to use this correlation curve obtained from such a few data to estimate the crystallite size. It is needed to accumulate the data to derive the accurate correlation curve. Fig.14 shows the result of comparison the value of G-FWHM with crystallite size La obtained by XRD. The value of G-FWHM of polished graphite is on the same correlation line as other carbon materials. Therefore, it is possible to use the value of G-FWHM of polished graphite to estimate graphite crystal size. It also indicates that the structural damage of graphite caused by polishing is supposed to be low. The relations between the ratio of D-band to G-band intensity and the crystallite size La was given by

$$\frac{10^3}{1.22 La} + 14.56 = \Delta 1580$$

Crystal size distribution by Raman microscopy

The 532nm pulsed laser beam was focused to 1μm in diameter with the objective lens (x100) to obtain the distribution of crystal size in detail. The total of 150 spectra was obtained from each material. The crystal size was calculated by using the relations obtained.

Fig.15 and Fig.16 shows the results of calculation for crystal size distribution. G347S has wider size distribution than that of G458S. The average crystal size of G347S is slightly larger than that of G458S. Each crystal size is 120nm (G347S) and 108nm (G458S), respectively. These average values are reasonable compared to bulk properties obtained by XRD.

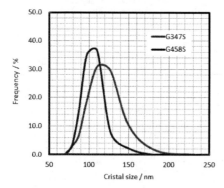
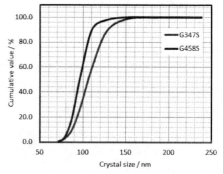

Fig.15 Crystal size distribution   Fig.16 Cumulative value

(3)Estimation of anisotropy

Developed crystal of graphite shows characteristic polarized properties depends on layered and planar structure (Fig.17) [1]. Therefore the polarized properties provide the information of anisotropy and the degree of graphitization. The mapping images were produced by analysis of reflected light from graphite materials using polarized light. The 577nm pulsed laser beam was used to estimate the anisotropy of graphite. Total illuminated area is 0.25mm^2 (0.5x0.5mm) and the pixel size of this study is 5μm. Therefore the total numbers of pixels are 10201. The angle of polarized light was changed from 0° to 180°.

Fig.17 Polarized properties of graphite

Fig.18 shows the mapping image of the ratio of maximum and minimum reflection intensity. This mapping image attributes to the degree of graphitization. The color of individual pixels ranges between blue (low graphitization) and red (high graphitization). Fig.19 shows the mapping image of the angle of maximum reflection intensity of G347S. This mapping image attributes to the direction of orientation. The color of individual pixels ranges between blue (0°) and red (180°). It is possible to estimate the degree of graphitization and its direction of orientation simultaneously, and the isotropic properties of G347S are confirmed visually. On the other hand, the resolution of this study is low compared to particle size consisting graphite. Therefore the further study with high resolution is needed to evaluate in detail. The data obtained by this method plans to be changed into a histogram and evaluated the distribution.

Fig. 18 Mapping image of the ratio of maximum   Fig. 19 Mapping image of the angle of maximum
and minimum reflection intensity   reflection intensity

CONCLUDING REMARKS

Systematic characterization of the graphite microstructures before and after irradiation and correlate the structural changes with the evolving macroscopic properties is proceeding to develop graphite for improved radiation service life. This paper introduces one of the initial results of microstructural evaluation for the pre-irradiated materials G347S and G458S, which is fine-grained nuclear graphite manufactured by Tokai Carbon.

The pore size distribution was measured by optical image analysis method and mercury porosimetry. The accuracy of evaluation was examined and the usefulness of the direct analysis method was shown. The crystal size distribution was also measured with Raman microscopy. The relations between the ratio of D-band to G-band intensity about polished graphite and the crystallite size La was given and the crystal size distribution was obtained by using this relations. On the other hand, the evaluation of anisotropy using polarized light was needed further study because of the insufficiency of resolution.

FOOTNOTES

*1 Optical Microscope

*2 Mercury Porosimetry

*3 Optical Anisotropy Mapping

REFERENCES

[1] Eagle G.B., Carbon 9, 539 (1971)

[2] Eagle G.B., B.T.Kelly, J. Nucl. Mater. 122&123, 122 (1984)

[3] J.E.Brocklehurst, B.T.Kelly, Carbon 31, 155 (1993)

[4] B.T.Kelly, T.D.Burchell, Carbon 32, 499 (1994)

[5] C.Berre, S.L.Fok, B.J.Marsden, L.Babout, A.Hodgkins, T.J.Marrow, T.M.Mummery, J. Nucl. Mater. 352, 1 (2006)

[6] G.Hall, B.J.Marsden, S.L.Fok, J. Nucl. Mater. 353, 12 (2006)

[7] A.N.Jones, G.N.Hall, M.Joyce, A.Hodgkins, K.Wen, T.J.Marrow, B.J.Marsden, J. Nucl. Mater. 381, 152 (2008)

[8] J.Kane, C.Karthik, D.P.Butt, W.E. Windes, R.Ubic, J. Nucl. Mater. 415, 189 (2011)

[9] R.J.Nemanich, S.A.Solin, Phys. Rev. B20, 392 (1979)

[10] F.Tuinstra, J.L.Koenig, J. Chem. Phys. vol.53, 1126 (1970)

[11] K.Nakamura, M.Fujitsuka, M.Kitajima, Chim. Phys. Lett. 172, 205 (1990)

[12] G.Compagnini, O.Puglisi G.Foti, Carbon 35, 1793 (1997)

[13] A.C.Ferrari, J.Robertson, Phys. Rev. B61, 14095 (2000)

[14] G.Katagiri, H.Ishida, A.Ishitani, Carbon 26, 565 (1988)

A MODEL FOR SIMULATION OF COUPLED MICROSTRUCTURAL AND COMPOSITIONAL EVOLUTION

Veena Tikare, Eric R. Homer and Elizabeth A. Holm
Sandia National Laboratories
PO Box 5800, MS 0747
Albuquerque, NM 87185-5800
VTikare@sandia.gov, ehomer@sandia.gov, eaholm@sandia.gov

ABSTRACT

The formation, transport and segregation of components in nuclear fuels fundamentally control their behavior, performance, longevity and safety. Most nuclear fuels enter service with a uniform composition consisting of a single phase with two or three components. Fission products form introducing more components. The segregation and transport of the components is complicated by the underlying microstructure consisting of grains, pores, bubbles and more, which is evolving during service. As they evolve, components and microstructural features interact such that composition affects microstructure and vice versa. The ability to predict compositional and microstructural evolution in 3D as a function of burn-up would greatly improve the ability to design safe, high burn-up nuclear fuels.

We present a model that combines elements of Potts Monte Carlo, MC, and the phase-field model to treat coupled microstructural-compositional evolution. The evolution process demonstrated is grain growth and diffusion in a two-phase system. The hybrid model uses an equation of state, EOS, based on the microstructural state and composition. The microstructural portion uses the traditional MC EOS and the compositional portion uses the phase-field EOS:

$$E_{hyb} = \sum_{i=1}^{N}\left(E_v(q_i,C) + \frac{1}{2}\sum_{j=1}^{n} J(q_i,q_j) \right) + \int \kappa_c (\nabla C)^2 dV$$

E_v is the bulk free energy of each site i and J is the neighbor interaction energy between neighboring sites i and j. The last term is the compositional interfacial energy as defined in the traditional phase-field model. The coupled microstructure-composition fields evolve by minimizing the free energy in a path dependent manner. An application of this modeling framework demonstrates the expected microstructural and phase coarsening, which is controlled by long-range diffusion.

INTRODUCTION

Current fuel performance codes address engineering performance by empirically modeling thermal conductivity, fission gas release, strain due to swelling and other physical processes. They do not concern themselves with the underlying compositional changes due to changing chemistry or microstructural changes that cause these phenomena. Thus, their predictive ability is limited. Well-developed models exist for simulating microstructural evolution by processes such as grain growth, recrystallization, Ostwald ripening and sintering as well as for

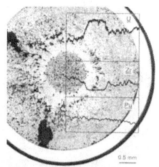

Figure 1. Component segregation in U-Pu-Zr metallic fuel accompanied by regions of very different microstructural evolution[1].

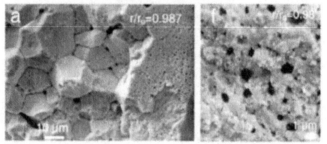

Figure 2a. Dynamic recrystallization at the outer rim of fuels containing [238]U leads to nano-sized grains only at the rim as shown above[3].

1.

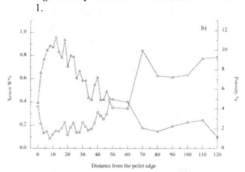

Figure 2b. Accompanying the recrystallization in 2a are changes in porosity and Xe distribution in the rim[3]. Simulating recrystallization, pores reconfiguration with fission gas distribution will require a coupled microstructure-composition evolution model.

compositional evolution by diffusion and segregation. However, models coupling microstructure and composition have been lagging. This coupling would enable simulation and prediction of processes such as component segregation in metal fuels[1,2] (shown in Figure 1), dynamic recrystallization and fission product distribution in light water reactor (LWR) fuels as a function of radial distance and in the rim region[3,4] (shown in Figure 2), diffusion of fission products, coarsening of grains and gas bubble in the kernel and chemical attack of the outer layers in tri-structural isotropic (TRISO) fuels[5] (shown in Figure 3), chemical interaction between the fuel and clad, restructuring of fast reactor fuels and many other microstructural-compositional evolution processes seen in all types of nuclear fuels. In all these processes, the compositional and microstructural evolution are coupled, meaning that the changes in one influences the other and therefore must be captured in the same model. Treating them as separate and parallel processes does not capture its interdependence and therefore cannot be predictive. The composition, microstructure and their evolution during service controls the engineering performance, safety and longevity of fuels during normal and off-normal operation and post-service storage. Should the spent fuel be reprocessed, the microstructure and composition may have implications for the particular process used for separation. Thus controlling and predicting such behavior is important.

The two primary models currently used for microstructural models are the Potts Monte Carlo (MC) method and the phase-field model. Potts MC is a statistical-mechanical model that populates a lattice with an ensemble of discrete particles to represent and evolve the microstructure. The particles evolve in a variety of ways to simulate microstructural changes. Potts MC methods have proven themselves to be versatile, robust and capable of simulating various microstructural evolution processes. They have the great advantage of being simple and intuitive, while still being a rigorous method that can incorporate all the thermodynamic, kinetic and topological characteristics to simulate complex processes. They are easy to code, readily extendable from 2D to 3D and can simulate the underlying physics of many materials evolution processes based on the statistical-mechanical nature of the model. These processes include curvature-driven grain growth[6,7], anisotropic grain growth[8], recrystallization[9], grain growth in the presence of a pinning phase[10,11], Ostwald ripening[12,13], and particle sintering[14,15,16,17]. The equation of state characterizing the materials in Potts MC is the sum of the bulk energy of each particle at each site i and the sum of all the interfacial energy of each particle as

$$E_{MC} = \sum_{i=1}^{N} \left(E_v(q_i) + \frac{1}{2} \sum_{j=1}^{n} J(q_i, q_j) \right) \tag{1}$$

where N is the total number of particles, E_v is the bulk energy the particle at site i, J is the neighbor interaction energy of particle at site i with its neighbor j for a total number of neighbors n and q_i is the grain orientation and or phase of particle at site i. Highly tailored equations of state for many different types of materials processes can be constructed using this basic equation.

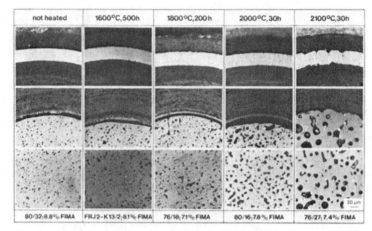

Figure 3. Fission gas bubble formation and coarsening in TRISO fuel kernels with diffusion of fission products from the kernel into outer layers. Fission product-layer chemical interactions can degrade outer layers[5].

In contrast, the phase-field model is a deterministic, continuum, thermodynamic model that describes the microstructure and its evolution in terms of continuum "phase-fields" that are evolving. The phase-fields can be grain orientations, composition, phases or other microstructural and compositional descriptors. While phase-fields are continuum quantities, their evolution is simulated by solving their field equations on some predefined grid, typically square in 2D and cubic in 3D. The size of the grid is such that the distance between solution points is much larger than atoms, but sufficiently small to resolve microstructural features such as grain boundaries. The phase-field method is a powerful mesoscale modeling method that is widely used to model the kinetics of microstructural evolution in materials. The appeal of the phase-field method is that it uses directly constructed free energy functionals as functions of the "phase-fields" to describe their thermodynamics (equation of state) and use the time-dependent Ginzburg-Landau equation to treat the evolution of the non-conserved phase-fields and the Cahn-Hilliard equation for conserved phase-fields. The phase field model has been used extensively to simulate many microstructural evolution processes including grain growth[18] and Ostwald ripening[19], gas bubbles in nuclear fuels[20,21], void ensembles under irradiation[22], precipitate morphology and evolution in alloys[23], nucleation and growth near a dislocation[24], coarsening of precipitates[25] and solidification[26,27]. The phase-field equation of state is a sum of the bulk energy and interfacial energies as a function of all Q phase-fields η_i integrated over the simulation volume V as:

$$E_{pf} = \int \left(f_o + \sum_{i=1}^{Q} \kappa_{\eta i} (\nabla \eta_i)^2 \right) dV \qquad (2)$$

where f_o is the bulk free energy and is a function of all the phase fields Q and their interaction with each other. An example of a free energy functional used for grain growth is:

$$f_o = \sum_{i=1}^{Q}\left(-\alpha\eta_i^2 + \beta\eta_i^4\right) + \gamma\sum_{i=1}^{Q}\sum_{j\neq i}^{Q}\eta_i^2\eta_j^2 \tag{3}$$

where α, β, γ are phenomenological constants used to obtain the desired thermodynamics. The second term with $\kappa(\nabla\eta_i)$ is the interfacial energy term where κ is a phenomenological constant used to define the energy and width of the interface. Like the Potts MC model, highly tailored equations of state for many different types of materials processes can be constructed using this basic equation.

However, both models do have drawbacks. Since, the Potts MC model uses an ensemble of particles to represent microstructure, smoothly varying compositions such as those required for the examples given in Figures 1 to 3 are difficult to represent and require very large simulations. Furthermore, diffusive transport has been simulated by random walk of particles; again requiring large ensembles for accurately sampling gradients in composition. Phase-field models are able to represent compositional gradients overlaying microstructural features well by using different phase-fields, but construction of the free energy functionals E_{pf} for coupling these phase-fields is very difficult and introduction of each set of couple phase-fields reduces the time and spatial increments to resolve them. Thus, simulations with coupled phase-fields become very time consuming to develop the free energy functionals and very computationally intensive, requiring massively parallel simulations. Almost all phase-field simulations published in the literature apply the technique to small 2D simulations to demonstrate the technique rather than to production-scale materials process simulation that can be used to study mesoscale materials behavior.

2. MODEL

To enable true mesoscale simulation of coupled compositional-microstructural evolution to characterize and study materials behavior, we propose to develop hybrid Potts MC-phase-field techniques that capture the robustness of Potts MC models while preserving the continuum compositional gradient terms of the phase-field model. We will use both a particle ensemble with statistical-mechanics and a continuum phase-field to capture the microstructural features and compositional variations, respectively. The microstructural representation is shown in Figure 4. A square grid is used in 2D to represent the microstructure where each site of the grid i has an integer value q_i (represented by different colors in Figure 4a) that represents its phase state and membership in a grain. Contiguous sites with the same state value form a grain of the associated phases. In these simulations, a total of 200 different integer values are used to identify grains, with $q_i = 1$ to 100 are grains of one phase, the α-phase, and $q_i = 101$ to 200 are of the other phase, β-phase. In addition, the same square grid has a phase-field value, which represents the composition as a continuum field. While the composition is a continually varying field as shown in Figure 4b, discrete compositional values C_i at a point in each site i are tracked and evolved numerically to simulate diffusion.

The equation of state for this hybrid model will become

$$E_{hyb} = \sum_{i=1}^{N} \left(E_v(q_i,C) + \frac{1}{2}\sum_{j=1}^{n} J(q_i,q_j) \right) + E_{dC} \qquad (4)$$

with the bulk energy term E_v as a function of both the microstructural state q_i and of composition, C. Plus the term $E_{dC} = \int 2\kappa_c (\nabla C)^2 dV$ that is only a function of the compositional gradients. The coupling is handled entirely by the $E_v(q_i,C)$ and κ_c terms and is much simpler than the phase-field case of coupling all phase-fields including all Q orientation phase-fields. In this representation, there is only one state locally and therefore the coupling is only between two variables rather than between all the phase-fields of each state that can be possible in the simulation.

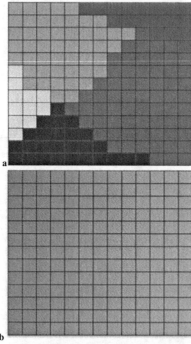

Figure 4. Microstructural and compositional representation used in the hybrid model. The microstructure (a) is overlaid on the same grid as the composition (b). The color in (a) is discrete and represents membership in a microstructural feature such as a grain; in (b) it represents the local composition, which is a continuum value and is smoothly varying.

Consider a simple two-component, two-phase system. The components are A and B; phases are α and β. At some temperature T, the bulk free energy of this system for the two phases as a function of component composition is:

$$E_v = \gamma\left[\left(C - C_1\right)^2 + \left(C_2 - C\right)^2\right] + a_1\left(C - C_3\right)^2 q_\alpha + a_2\left(C_4 - C\right)^2 q_\beta \qquad (5)$$

where C is the phase-field representing the composition and γ, a_1, a_2 C_1, C_2, C_3 and C_4 are phenomenological constants chosen to obtain the desired free energy for each phase as a function of composition. The free energies of phases α and β given by eq. (5) as a function of composition for $\gamma = 0.3$, $a_1 = a_2 = 0.5$, $C_1 = 0.25$, $C_2 = 0.75$, $C_3 = 0.05$ and $C_4 = 0.95$ are plotted in Figure 5. Note $q_\alpha = 1$ and $q_\beta = 0$ for the α-phase and $q_\alpha = 0$ and $q_\beta = 1$ for the β-phase in eq (5). For this formulation of bulk free energy, the α-phase is the stable phase from composition $C_A = 0$ to 0.3 and β is stable from $C_A = 0.7$ to 1.0; in between, a mixture of the two phases is stable.

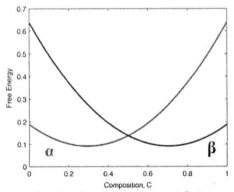

Figure 5. Free energy as a function of composition for α- and β-phases as defined by eq (4) over the composition ranging from 0 to 1.0 fraction B in the A-B component system.

The microstructural-compositional evolution is handled by iteratively treating the particle ensemble and the compositional field with the coupling shown above. The microstructural evolution by grain growth of each phase is handled as it is in Potts MC model and is described in detail in other works[4]. For each Monte Carlo time step, MCS, every site in the simulation space attempts a change in its microstructural state. Each site i attempts changing it grain state q_i to that of one of its neighbors chosen at random. The change in grain state q_i is not restricted to the same phase; it is free to change to the other phase. The change in energy for such a change is evaluated using eq (4) with the volume free energy E_v given by eq (5) and the interfacial energy component is $E_i^{gb} = \sum_{j=1}^{n} J(q_i q_j)$ where $J(q_i,q_j) = 1$ when $q_i \neq q_j$ and $J(q_i,q_j) = 0$ when $q_i = q_j$. Note, that the volume free energy portion contributes only should the site attempt a phase change. If the attempted grain state change is to a different phase, then the local composition at that site influences the change in energy associated with that change. The change is carried out using the Metropolis algorithm. The probability of the change P is calculated using Boltzmann statistics with the probability of change is a function of the change in energy. The change in energy is $\Delta E = E_{final} - E_{initial}$ calculated using eq (4).

$$P = \exp\left(-\frac{\Delta E}{k_B T}\right) \quad for \quad \Delta E > 0$$
$$P = 1 \quad for \quad \Delta E \leq 0$$

(6)

If the probability $P = 1$, the change is carried out. If $P < 1$, then a random number R that is evenly distributed from 0 to 1 is chosen. If $R \leq P$, then the site is changed to it new state. The simulation temperature used in this work is $k_B T = 0.3$.

After one MCS where all the sites have attempted a microstructural change, the composition phase field is updated to simulate diffusion by using the Cahn-Hilliard equation

$$\frac{\partial C}{\partial t} = M_c \left(\nabla^2 \frac{\partial E_v}{\partial C} - \kappa_c \nabla^4 C \right)$$

(7)

where M_c is the mobility and is related to the diffusion coefficient and κ_c gradient term. At each site i, composition C_i is updated using eq (7) with standard numerical techniques for spatial differentiation of composition and $C(i,t+1) = \left.\frac{\partial C}{\partial t}\right|_{i,t} \Delta t_{pf} + C(i,t)$. The time increment used for this is $\Delta t_{pf} = (1/10)$ MCS; in other words, we performed 10 iterations on the composition field for every MCS. In the simulations presented in this work, $M_c = 1.0$ and $\kappa_c = 1.0$.

The overall phase composition of the simulation (i.e. the total number of sites of the $\alpha-$ and β-phase) is not conserved numerically. The only constraint on phase and grain changes attempted by the Potts MC portion is introduced through the energetics of the system. It is energetically favorable for α–sites to have compositions $C = 0.0$ to 0.5 with the lowest energy corresponding to $C_\alpha = 0.3$ and β–sites to be of $C = 0.5$ to 1.0 with the lowest energy corresponding to $C_\beta = 0.7$. Unlike the phase composition, the overall chemical composition is conserved by the Cahn-Hilliard eq (7) and remains constant through out the simulation.

SIMULATION RESULTS AND DISCUSSION

A two-phase, two-component system, using the model described in the previous section, was simulated with grain growth, phase growth and composition segregation all occurring simultaneously. The starting condition for the simulation was an ensemble of particles consisting of approximately equal number of α- and β-particles populating the lattice in random positions. The starting composition and microstructure are tiny grains of α- and β-particles with their compositions set to $C = 0.30$ and 0.70 with an approximate overall composition of $C = 0.50$. The bulk free energies of the two phases, α and β, as a function of composition given by eq. (5) are plotted in figure 5. The interfacial free energy of these tiny grains is the contribution from the last two terms in eq (4), the grain boundary (sum of bonds between unlike neighboring particles) and the gradient in concentration, $2\kappa_c(\nabla C)^2$. Thus the starting configuration has high interfacial energy as the grain boundary length is extensive and the gradient in composition at these boundaries is large. Examination of figure 5 shows that in a system described by these thermodynamic characteristics, the α-phase will have it lowest bulk free at composition $C = 0.3$ and β-phase at composition $C = 0.7$. Grains of both phases will lower their grain boundary energies by grain growth and minimizing interphase energies by coarsening of phase regions. Simulation results show that this is indeed the case.

Figure 6 shows the evolution of grains, phase and composition for such a system. Grains of both phases are growing during the simulations. The phase regions (clusters of grains of the same phase) are also growing. These results are consistent with expected behavior as the interfacial free energy is minimized by coarsening. This phase coarsening is accompanied by changes in composition by inter-diffusion of the two components so that the α-regions are composition $C \approx 0.3$ and β-region are $C \approx 0.7$. This again is consistent with expected results as the overall interfacial free energy is minimized and the bulk energy of each phase is lowest when it is at its equilibrium composition. While grain growth occurs by short-range diffusion, the phase coarsening and compositional evolution occur by long-range diffusion. In a two-phase system with each phase having a different composition such the one simulated here, grain growth, phase coarsening and component redistribution all occur in lock step, with the inter-diffusion of components A and B being the rate limiting process. There is higher solubility of the minor component at the phase edges and in features with higher curvature, consistent with predictions for these systems.

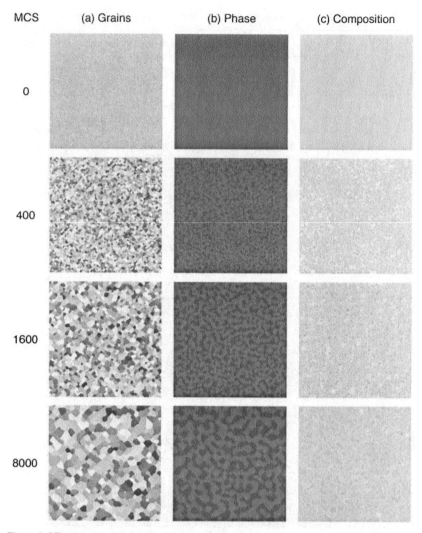

Figure 6. Microstructural and compositional evolution is shown as (a) grain growth, (b) phase-region coarsening and (c) compositional evolution at t = 0, 500, 1600 and 8000 MCS.

The hybrid model, which combines elements of Potts MC model with the phase-field model, is described and demonstrated by applying to coupled grain growth and diffusion in a two-phase system. The hybrid model is much more efficient than a phase-field as the number of "phase-fields" is reduced from $Q + 1$ to 1. Q, which is typically $Q = 100$, is the total number of grain orientations and the additional phase-field describes the composition. The evolution of compositional phase field treated by eq (7) is more efficient as the complexity of the $\frac{\partial E_v}{\partial C}$ term is far lower than for a proper phase-field simulation[28,29]. Lastly, the construction of a free energy functional describing the system thermodynamics is greatly simplified as the coupling of the composition to phase-fields describing grains orientation and phase have been eliminated. The coupling is now directly to the single Potts model parameter q_i that describes the local grain orientation and phase.

Similarly, the hybrid model enables simulation of materials phenomena that would be difficult and inefficient in the classical Potts MC model. While microstructural evolution controlled by long-range diffusion has been simulated using the Potts MC model[12,11], the diffusion is simulated by random walk of individual particles through the digitized lattice. This limits the applicability of Potts MC methods to materials systems where the compositional gradients across the different microstructural features need not be considered in great detail. Alternatively, very large Potts MC simulations with large numbers of walkers can capture the necessary compositional gradients, however, the size of such a simulation is prohibitive. Thus, this hybrid model combines the capabilities to enable a new set of phenomena that of great interest to nuclear materials scientists.

CONCLUSION

A novel hybrid model that combines the Potts MC and phase-field models has been introduced in the work. The model is demonstrated by applying it to a two-component, two-phase system that forms a simple eutectic. Coupled grain growth, phase coarsening and compositional evolution by long-range diffusion have been demonstrated. This hybrid model not only simulates these processes correctly, it is easier and simpler to implement and numerically more efficient than either the Potts MC or phase-field models for materials evolution processes involving long-range diffusion through complex microstructures. This new capability will prove very useful for many nuclear materials and chemistry problems pertaining to fuel, clad, core and waste disposal materials.

ACKNOWLEDGMENTS

Sandia National Laboratories is a multi-program laboratory managed and operated by Sandia Corporation, a wholly owned subsidiary of Lockheed Martin Corporation, for the U.S. Department of Energy's National Nuclear Security Administration under contract DE-AC04-94AL85000

[1] Y.S. Kim G.L Hofman, S.L. Hayes and Y.H. Sohn, "Constituent redistribution in U-Pu-Zr fuel during irradiation" *J. Nucl. Mat.*, 2004, 327 27

[2] Y.H. Sohn, M.A. Dayananda, G.L. Hofman, R.V. Strain and S.L. Hayes, "Analysis of constituent redistribution in the gamma (bcc) U-Pu-Zr alloys under gradients of temperature and concentrations", *J. Nucl. Mat.*, 2000, 279 317

[3] J. Spino and D. Papaioannou, "Lattice contraction in the rim zone as controlled by recrystallization: Additional evidence", *J. Nucl. Mater.*, 2008, 372 416-420

[4] N. Lozano, L, Desgranges, D. Aymes and J.C. Niepce, "High magnification SEM observations for two types of granularity in a high burnup PWR fuel rim", *J. Nucl. Mater.*, 1998, 257, 78-87

[5] H. Nabielek, W. Kuhnlein, W. Schenk, W. Heit. A. Christ and H. Ragoss, "Development of advanced HTR fuel-elements", *Nucl. Eng. & Desg.*, 1990, 121 199-210.

[6] M. P. Anderson, D. J. Srolovitz, G. S. Grest, and P. S. Sahni, Computer-simulation of grain-growth. 1. Kinetics" *Acta Metall.*, 1984, 32, 783-791

[7] E. A. Holm, J. A. Glazier, D. J. Srolovitz, G. S. Grest, "Effects of lattice anisotropy and temperature on domain growth in the 2-dimenstional Potts-model", *Phys. Rev. A*, 1991, 43, 2662-2668.

[8] E. A. Holm, G. N. Hassold and M. A. Miodownik, "On misorientation distribution evolution during anisotropic grain growth", *Acta Mater.*, 2001, 49, 2981-2991

[9] E. A. Holm, M. A. Miodownik and A. D. Rollett, "On abnormal subgrain growth and the origin of recrystallization nuclei, *Acta Mater.*, 2003, 51, 2701-2716

[10] M. Miodownik, J.W. Martin and A. Cerezo, "Mesoscale simulations of particle pinning", *Phil. Mag. A*, 1999, 79, 203-222

[11] V. Tikare, M. Miodownik, E. A. Holm, "Three Dimensional Simulation of Grain Growth in the Presence of Mobile Pores," *J. Am. Ceram. Soc.*, 2001, 84, 1379-1385

[12] V. Tikare and J.D. Cawley, "Numerical Simulation of Grain Growth in Liquid Phase Sintered Materials – I. Model," *Acta Mater.* 1998, 46 1333-1342

[13] V. Tikare and J. D. Cawley, "Numerical Simulation of Grain Growth in Liquid Phase Sintered Materials – II. Study of Isotropic Grain Growth," *Acta Mater.*, 1998, 46, 1343-1356.

[14] G.N. Hassold, I-W. Chen, D. J. Srolovitz, *J. Am. Ceram. Soc.,* 1990, 73, 2857-2864

[15] M. Braginsky, V. Tikare and E. Olevsky, "Numerical Simulation of Solid State Sintering," *Int. J. Sol. Struc.,* 2004, 42, 621-636

[16] V. Tikare, M. Braginsky and E.A. Olevsky, "Numerical Simulation of Solid-State Sintering: I, Sintering of Three Particles," *J. Am. Ceram. Soc.*, 2003, 86, 49-53

[17] V. Tikare, M. Braginsky, D. Bouvard and A. Vagnon, "Numerical Simulation of Microstructural Evolution during Sintering at the Mesoscale in a 3D Powder Compact," *Comp. Mat. Sci.*, 2010, 48, 317-325

[18] D. Fan and L. Q. Chen, "Computer simulation of grain growth using a continuum field model", *Acta Materialia*, 1997. 45 611-622

[19] V. Tikare, E.A. Holm, D. Fan and L.Q. Chen, "Comparison of Phase-Field and Potts Models for Coarsening Processes" *Acta Mater.* 47 363-371 (1999).

[20] S. Y. Hu, C. H. Henager Jr., H. L. Heinisch, M. Stan, M. I. Baskes and S. M. Valone, "Phase-field modeling of gas bubbles and thermal conductivity evolution in nuclear fuels", *J. Nucl. Mater.,* 2009, 392, 292-300

[21] S. Y. Hu, Y. L. Li, X. Sun, F. Gao, R. Devanathan, C. H. Henager Jr. and M. Khaleel, "Application of the phase-field method in predicting gas bubble microstructure evolution in nuclear fuels", *Intl. J. Mater. Res.,* 2010, 101 (4), 515-522

[22] S. Y. Hu and C. H. Henager Jr., "Phase-field modeling of void lattice formation under irradiation", *J. Nucl. Mater.,* 2009, 394, 155-159

[23] S. Y. Hu and C. H. Henager Jr., "Phase-field simulations of Te-precipitate morphology and evolution kinetics in Te-rich CdTe crystals", *J. Crystal Growth,* 2009, 311, 3184-3194.

[24] S. Y. Hu and L. Q. Chen, "Solute segregation and coherent nucleation and growth near a dislocation - A phase-field model integrating defect and phase microstructures", *Acta Mater.*, 2001, 49, 463-472.

[25] V. Vaithyanathan and L. Q. Chen, "Coarsening of ordered intermetallic precipitates with coherency stress", *Acta Mater.*, 2002, 50, 4061-4073.

[26] W. J. Boettinger, J. A.Warren, C. Beckermann, and A. Karma, "Phase-field simulation of solidification", *Annu. Rev. Mater. Res.*,2002, 32, 163-194.

[27] L. Q. Chen, "Phase-field models for microstructure evolution", *Annu. Rev. Mater. Res*, 2002, 32, 113-140

[28] D. Fan and L-Q. Chen, "Diffusion-controlled grain growth in two-phase solids", *Acta Metal.*, 1997, 45[8] 3297-3310

[29] T.P. Swiler, V. Tikare and E.A. Holm, "Heterogeneous diffusion effects in polycrystalline microstructures", *Mat. Sci. and Eng.*, 1997, A238 85-93

CHARACTERISATION OF CORROSION OF NUCLEAR METAL WASTES ENCAPSULATED IN MAGNESIUM SILICATE HYDRATE (MSH) CEMENT

Tingting Zhang, Chris Cheeseman
Department of Civil & Environmental Engineering, Imperial College London
South Kensington Campus, SW7 2AZ London, UK

Luc J. Vandeperre
Department of Materials & Centre for Advanced Structural Ceramics, Imperial College London
South Kensington Campus, SW7 2AZ London, UK

ABSTRACT

A novel low pH magnesium silicate hydrate cement system for encapsulating nuclear industry wastes have been developed using blends of MgO, silica fume (SF), $MgCO_3$ and sand. Aluminium and Magnox swarf were encapsulated in both this new system and in a BFS/PC control system used in the nuclear industry. The interaction of the optimised mortar with the metal strips has been investigated, both in terms of rate of continued corrosion as well as the phases that form by reaction of the binder with different metal strips. Magnox swarf was better bound into the BFS/PC system than MgO/SF system whereas Al 1050 metal strips were bound better into the MgO/SF samples than into the BFS/PC reference mortar. No H_2 generation was recorded when aluminium or magnox were encapsulated in the new binder, which is substantially better than what can be achieved with the reference system. Hence, the newly developed binder could potentially encapsulate mixtures of reactive metals better than the existing solution.

INTRODUCTION

Nuclear wastes are mainly produced in nuclear power stations and are generated from the processing and associated production of nuclear fuel. There is a considerable amount of intermediate level legacy nuclear waste in the UK which varies in terms of physical form and chemical composition. The inhomogeneous nature of these wastes is a major challenge to formulate reliable and robust systems for waste encapsulation[1].

Encapsulation in cements has been shown to be a viable option for some of the wastes and a good record of treatment and containment for a range of wastes has been accrued over the years[2]. Composite cements based on the partial replacement of ordinary Portland cement (OPC) with blastfurnace slag (BFS) or pulverised fuel ash (PFA) are commonly used in the UK, but research into potential alternatives has evaluated the potential for waste encapsulation using calcium aluminate cement, magnesium phosphate cement, calcium phosphate cement, calcium sulpho-aluminate cement, alkali-activated systems and geo-polymers[3].

Aluminium and Magnox, a magnesium alloy, are the main metal wastes in the legacy intermediate level radioactive waste (ILW) stream in the UK. They arise from the de-cladding of nuclear fuel and components of the fuel rods. In the legacy waste stream, accrued during past operations, these metals have not been separated. This poses a challenge for treatment since the requirements for passivation of corrosion of these metals are very different. A moderate pH (pH 4-10) is advantageous for amphoteric aluminium because both high pH and low pH cause corrosion[4]. Additionally, at high pH the corrosion also generates H_2 through:

$$2Al + 2OH^- + 6H_2O \rightarrow 2[Al(OH)_4]^- + 3H_2(g) \uparrow \qquad (1)$$

In sharp contrast, passivation of magnesium requires a high pH[4] as the solubility of magnesium hydroxide decreases with pH.

Considering that aluminium may still be passive for pH values up to 10 and that magnesium hydroxide could remain stable down to pH 9, it was decided to design cements whose pore water pH would be in the range of pH 9-10. An expected further advantage of this approach is that a much wider range of metals shows low corrosion and/or mobility in this pH range[4]. Hence, such a binder could potentially have much wider application than in legacy waste encapsulation, which would enhance the commercial viability of the product. Earlier work has shown that the reaction between magnesium oxide (MgO) and silica fume (SF) yields magnesium-silicate-hydrate gel (MSH), and that this cement has a good strength and a pH in the desired range[5]. It was shown also that partial substitution of MgO with magnesium carbonate ($MgCO_3$) allows buffering the initial pH of the system around pH 10 and that addition of sand to produce a mortar rather than a neat cement improves the dimensional stability[6].

In this work the corrosion of metal wastes in the optimised M-S-H binder will be compared to the corrosion in one of the current industry standard cements (OPC/PFA). A first set of measurements details the hydrogen gas generation over time. This is followed by a characterisation of the interaction products that form using X-ray diffraction and scanning electron microscopy observation of sections showing the metal-cement interface.

EXPERIMENTAL

The raw materials used were Portland cement (PC; CEM II, Lafarge, UK), a commercially available magnesium oxide (MgO, MagChem 30, M.A.F. Magnesite B.V.), ground granulated blast furnace slag (BFS, Civil and Marine Slag Ltd., UK), silica fume (SF; Elkem Materials Ltd), magnesium carbonate ($MgCO_3$, Fisher, UK) and sand (RH110, Sibelco, UK). Metal wastes used are a high purity Al alloy (Al 1050) and magnox swarf.

Experiments were completed using a control composite cement consisting of 25 wt% OPC and 75 wt% BFS and a water to solids ratio of 0.33, and a M-S-H cement formulation containing 20 wt% MgO, 5wt% $MgCO_3$, 25 wt% SF and 50 wt% sand. The water to solids ratio was 0.35 and 1% of hexametaphosphate was added to improve the rheology of the cement. To establish that the binder systems used have sufficient strength for handling and storage, the unconfined compressive strength (UCS) of the samples was determined using a standard compression tester (Zwick/Roell). The load rate used was 300 kPa/s and the maximum load applied to the blocks was recorded. The interaction between metals and the binders was studied by encapsulating small strips of metal (25 mm × 6 mm × 3 mm) in the cement systems. The volume of H_2 generated from the samples was determined using the experimental setup shown in Figure 1.

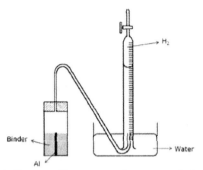

Figure 1. Schematic diagram of the set-up used to measure the hydrogen generation

X-ray diffraction (PW 1700 with Cu Kα radiation, Philips, The Netherlands) was used to identify the crystalline phases formed on the metal surface after 28 days of encapsulation. The variation of the chemical composition near the metal-binder interface was analysed by energy dispersive elemental analysis in a scanning electron microscope (JEOL-JSM-840A, Jeol, Japan). Samples were polished to a 1 μm surface finish, dried and gold coated before being examined.

RESULTS AND DISCUSSION

The development of the compressive strength of both types of binders is shown in Figure 2. The 28 day compressive strength of both mixes is similar and above 60 MPa, which is much higher than the strength required for nuclear waste encapsulation (>5 MPa)[7]. After 90 days curing, the optimised MgO/SF mix offers a higher compressive strength than PC/BFS mix of around 80 MPa. The continued strength gain also indicates full reaction of the cement requires more than 28 days.

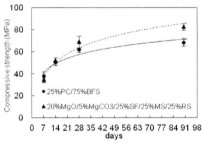

Figure 2. Compressive strength as a function of curing time for the reference and MgO/SF binders.

The volumes of hydrogen gas generated over time when aluminium is encapsulated in the two cements are compared in Figure 3. The initial rate of hydrogen evolution in the reference binder is extremely high, but this is followed by a marked slower rate of hydrogen evolution indicating that some form of passivating reaction layer may form. In the MgO/SF system, no H_2 evolution was detected during the test period. In a similar formulation without $MgCO_3$, the initial pH was found to be higher and hydrogen gas evolution was observed[6]. Therefore the suppression of corrosion in the current, optimized, binder is attributed to the buffering capacity of the magnesium carbonate.

As shown Figure 4, after 28 days of encapsulation, the Al strip in the reference binder (25% PC/ 75% BFS) has corroded significantly, whereas the surface of the Al strip embedded in the MgO/SF system shows no signs of surface attack. Moreover, the Al strip was found to be firmly bound to the binder.

Higher magnification images of the interface in both systems are shown in Figure 4. There is a clear reaction layer of up to 1 mm between the OPC/BFS cement and the Al strip, which also contains many large pores due to early virulent hydrogen evolution. In contrast, the interface between Al and MgO/SF is clean and devoid of clear reaction layers. The results of the electron probe micro-analysis, see Table 1, confirms this interpretation as the Al content near the interface is elevated in the OPC/BFS system whereas hardly any Al can be detected near the interface in the MgO/SF system.

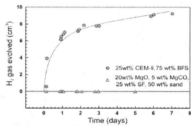

Figure 3. Hydrogen gas evolution versus time for aluminium encapsulation in a PC/BFS reference blend and in an MgO/SF binder.

(a)   (b)

Figure 4. Al sample after 28 days encapsulation in (a) PC/BFS, (b) MgO/SF

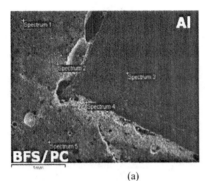 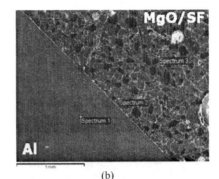

(a)   (b)

Figure 5. Interface between Al and the binder for (a) PC/BFS and (b) MgO-SF

Table 1. Electron probe micro-analysis results (wt%). Locations are indicated in the micrographs.

	O	Na	Mg	Al	Si	K	Ca
Al + PC-BFS							
Al							
Spectrum 3	7.58			85.33	6.96		0.12
Al / PC-BFS interface							
Spectrum 2	57.26	1.30	2.31	10.41	8.36	0.94	19.39
Spectrum 4	58.57	1.46	1.80	10.61	7.73	1.19	18.65
PC-BFS matrix							
Spectrum 1	51.70	0.84	3.49	4.59	11.96	1.03	26.33
Spectrum 5	50.59	0.87	3.61	4.82	13.15	1.03	25.93
Al + MgO-SF							
Al							
Spectrum 1	9.75			83.41	6.85		
Al/MgO-SF interface							
Spectrum 2	53.29	0.27	11.25	0.89	34.30		
MgO-SF matrix							
Spectrum 3	56.02		17.30	0	26.68		

The Al bars were then removed to allow examining the interface by XRD. The XRD patterns are presented in Figure 6 and show that the surface of the aluminium, which was encapsulated in BFS/PC system, consists mainly of Al(OH)$_3$. The aluminium removed from the optimized MgO/SF has hardly changed from normal aluminium with very weak peaks due to hydroxide products. This confirms again that there is hardly any corrosion of the Al bar when it is encapsulated in the MgO-SF binder.

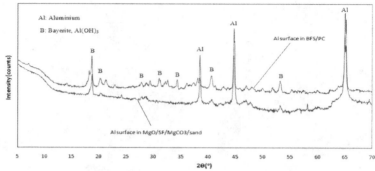

Figure 6. X-ray diffraction patterns of the surface of the aluminium removed from the binder after 28 days for OPC/BFS and the MgO/SF binder. Peaks have been labelled with Al for aluminium and B for Bayerite (Al(OH)$_3$).

Experiments with magnox swarf did not yield hydrogen generation at a level detectably by the simple set-up used for either binder, and as shown in Figure 7, the magnox swarf is bound into the binders quite well with no large gaps appearing for either binder consistent with the absence of hydrogen gas evolution at the level detectable with the set-up used. Higher magnification images obtained by scanning electron microscopy are shown in Figure 8. There is a limited interaction zone surrounding the magnox swarf in the OPC/BFS binder, which is confirmed by an elevated Mg content in the binder within 200 μm of the interface, see Table 2. The measurement in the bulk binder, spectrum 3, shows only 4.15 wt% Mg, which is consistent with what would be expected from the raw material data. Hence, encapsulation of magnox swarf in the reference binder does not lead to problems consistent with the successful industrial practice.

Surrounding the magnox in the MgO/SF binder, a gap has appeared. It is not entirely clear whether this is due to cracking during handling or whether the crack existed before the sample was prepared. A further change is that there is now a region with higher Mg and lower Si content (spectrum 2 and 4) suggesting that some magnesium from the magnox has diffused into the matrix. However, since magnesium is already a major component of the binder, this should not affect phase formation too much as it could merely lead to some brucite (Mg(OH)$_2$) formation in addition to M-S-H gel. Moreover, for this binder also the extent of the interaction layer is limited (< 200 μm).

(a)   (b)

Figure 7. Magnox swarf sample encapsulated in (a) PC/BFS, (b) MgO/SF

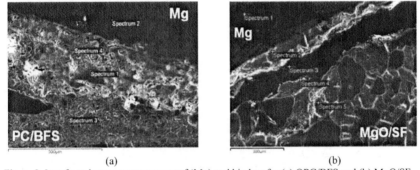

(a)   (b)

Figure 8. Interfaces between magnox swarf (Mg) and binders for (a) OPC/BFS and (b) MgO/SF

Table 2. Electron probe micro-analysis results (wt%). Locations are indicated in the micrographs.

		O	Na	Mg	Al	Si	K	Ca
Mg + PC-BFS								
Magnox								
	Spectrum 2	4.57		94.71	0.72			
interface								
	Spectrum 1	51.44		43.68	0.95	1.82	0.19	1.91
	Spectrum 4	58.07		24.99	1.48	3.65	0.29	11.10
PC-BFS matrix								
	Spectrum 3	49.40	0.85	4.15	4.62	13.60	0.78	11.10
Mg + MgO/SF								
Magnox								
	Spectrum 1	9.83		90.06				0.11
interface								
	Spectrum 2	56.43	0.34	20.01		22.35	0.40	0.48
	Spectrum 4	56.07	0.27	18.91		24.12	0.26	0.38
MgO-SF matrix								
	Spectrum 5	54.87		13.96		30.61	0.23	0.33

CONCLUSIONS

The compressive strength of both the MgO/SF binder and of the OPC/BFS binder after 28 days is higher than the required strength for nuclear waste encapsulation. The strength of MgO/SF system is similar to the control sample after 28 days curing but higher after 90 days, which suggest continued reaction after 28 days in the MgO/SF system.

No H_2 gas has been detected when Al is encapsulated in the MgO/SF binder, which can be explained by the lowering of the initial pH by substitution of MgO with $MgCO_3$. The surface of the Al strip embedded in the optimized MgO/SF system is clear and shows hardly any hydroxides and the Al strip was found to be firmly bound to the binder. In contrast Al encapsulated in the OPC/BFS reference binder corrodes badly leading to pore formation and a large interaction zone elevated in Al.

When magnox swarf is encapsulated in the binders, no H_2 gas generation was detected for either system. Based on the observation that reaction layers do form in either system, the sensitivity of the simple set-up was probably too low to detect its formation. Layers with enhanced Mg levels were found for both binders but these are limited to approximately 200 μm.

These results indicate that the newly proposed MgO/SF binder could potentially encapsulate mixtures magnox swarf and aluminium without extensive corrosion, which is better than what can be achieved with the current reference system.

ACKNOWLEDGEMENTS

The authors would like to thank the Engineering and Physical Science Research Council (EPSRC) for funding of the project through grant EP/F055412/1, and Ed Butcher from the National Nuclear Laboratory Ltd. for useful discussions and supplying the metals.

REFERENCES

1. Ojovan, M.I. and W.E. Lee, An introduction to nuclear waste immobilisation. 2005, London: Elsevier. 315.
2. Caldwell, R.J., et al. Characterisation of full-scale historic inactive cement-based intermediate level wasteforms. in *Stabilisation/Solidification Treatment and Remediation*. 2005. Cambridge: A.A. Balkema Publishers.
3. Milestone, N.B., Reactions in cement encapsulated nuclear wastes: need for toolbox of different cement types. *Advances in Applied Ceramics*, 2006. **105**(1): p. 13-20.
4. Deltombe, E., C. Vanleugenhaghe, and M. Pourbaix, Atlas of Electrochemical Equilibria in Aqueous Solutions. 1966, Oxford: Pergamon Press. 168-175.
5. Zhang, T., C.R. Cheeseman, and L.J. Vandeperre, Development of low pH cement systems forming magnesium silicate hydrate (M-S-H). *Cement and Concrete Research*, 2011. **41**(4): p. 439-442.
6. Zhang, T.T., L. Vandeperre, and C.R. Cheeseman, Bottom-up design of a cement for nuclear waste encapsulation. *Ceramic Engineering and Science Proceedings*, 2011. **32**(9): p. TBC.
7. Ojovan, M.I. and W.E. Lee, An introduction to nuclear waste immobilisation 2005, London: Elsevier.

IMPACT OF URANIUM AND THORIUM ON HIGH TiO$_2$ CONCENTRATION NUCLEAR WASTE GLASSES

Kevin M. Fox and Thomas B. Edwards
Savannah River National Laboratory
Aiken, SC, USA

ABSTRACT

This study focused on the potential impacts of the addition of Crystalline Silicotitanate (CST) and Monosodium Titanate (MST) from the Small Column Ion Exchange (SCIX) process on the Defense Waste Processing Facility (DWPF) glass waste form and the applicability of the DWPF process control models. MST from the Salt Waste Processing Facility (SWPF) is also considered in the study. The KT08-series of glasses was designed to evaluate any impacts of the inclusion of uranium and thorium in glasses containing the SCIX components. All but one of the study glasses were found to be amorphous by X-ray diffraction (XRD); one of the slowly cooled glasses contained a small amount of trevorite, which is typically found in DWPF-type glasses and had no practical impact on the durability of the glass. The measured Product Consistency Test (PCT) responses for the study glasses and the viscosities of the glasses were well predicted by the current DWPF models. No unexpected issues were encountered when uranium and thorium were added to the glasses with SCIX components.

INTRODUCTION

The Savannah River Site (SRS) Liquid Waste contractor had planned to begin a process referred to as Small Column Ion Exchange (SCIX) to disposition salt solution in fiscal year 2014 (these plans have been put on hold at present). In the first step of the process, salt solution retrieved from various waste tanks will be struck with Monosodium Titanate (MST) to remove key actinides and strontium. The salt solution will then be processed using Rotary Micro Filtration (RMF) to remove the MST and any insoluble solids. The MST and insoluble solids will accumulate on the bottom of the waste tank. The filtrate from RMF will be fed to ion exchange columns, also in the waste tank, to remove the ^{137}Cs using Crystalline Silicotitanate (CST) resin. The decontaminated salt solution from SCIX will be sent to the Saltstone Facility for immobilization in grout. The ^{137}Cs-laden CST resin will be sluiced and ground for particle size reduction, then sent to the Defense Waste Processing Facility (DWPF) for immobilization in glass. These processes are similar to the current disposition paths for streams associated with the Salt Waste Processing Facility (SWPF), which is under construction and will run concurrently with SCIX.

The MST and CST from the SCIX process will significantly increase the concentrations of Nb$_2$O$_5$, TiO$_2$, and ZrO$_2$ in the DWPF feed. Other constituents of MST and CST – Na$_2$O and SiO$_2$ – are already present in high concentrations in DWPF glass; thus their influences are well understood. The increased concentrations of Nb$_2$O$_5$, TiO$_2$, and ZrO$_2$ will likely have some impact on the properties and performance of the DWPF glass product. Properties such as the liquidus temperature, viscosity, and rate of melting of the glass may be impacted. The performance of the glass, particularly its chemical durability as it pertains to repository acceptance requirements, may also be impacted. The DWPF uses a set of semi-empirical and first-principles models referred to as the Product Composition Control System (PCCS)[1] to predict the properties and performance of a glass based on its composition since it is not practical to measure these attributes during processing. The objective of this study is to evaluate the impacts of the SCIX streams on the properties and performance of the DWPF glass product and on the applicability of the current process control models.

EXPERIMENTAL PROCEDURE

Compositions labeled as the KT08-series were selected to evaluate any impacts of the inclusion of uranium and thorium in glasses with the SCIX components. While the composition projections for the sludge batches with SCIX additions included uranium and thorium,[2] these components are typically removed from the glasses fabricated for the experimental studies[3-5] in order to minimize exposure to radioactivity. Several variability studies performed at the Savannah River National Laboratory (SRNL) in support of frit optimization for DWPF processing have shown that the properties of glasses fabricated with uranium and thorium are unlikely to differ significantly from those of their non-radioactive counterparts.[6-9] The KT08-series glasses were selected to further confirm these findings when the SCIX components are included, as well as to determine whether changes in the amounts of the non-radioactive components in the glass (as a function of the total glass composition) have any significant impacts on the properties or performance of the glass, including model applicability.

The basis for the KT08-series compositions was a series of projections of individual sludge batches incorporating the SCIX streams. Composition projections for sludge batches 8 through 17 were used,[10] and CST additions to the DWPF feed tank were projected at the accelerated DWPF processing rate of 75 Sludge Receipt and Adjustment Tank (SRAT) batches per year (including MST) with the SWPF streams added. The final SRAT batch composition for each sludge batch was used, since these cases represent the maximum concentrations of CST in the sludge. The resulting ten sludge composition projections are given in Table I. Each projection is identified by the relevant sludge batch and SRAT batch number.

Table I. Projected Compositions (wt %) of the Final SRAT Batches of Sludge Batches 8 through 17, Including SCIX Streams, Used to Develop the KT08 Glass Compositions.

Oxide	SB08-69	SB09-79	SB10-80	SB11-70	SB12-71	SB13-66	SB14-74	SB15-91	SB16-38	SB17-35
Al$_2$O$_3$	14.25	12.68	10.85	12.29	17.00	17.86	12.51	10.96	12.14	12.51
BaO	0.17	0.17	0.17	0.17	0.17	0.17	0.17	0.17	0.17	0.17
CaO	2.30	2.31	2.32	2.12	2.32	2.43	1.98	1.76	2.11	2.16
Ce$_2$O$_3$	0.70	0.70	0.62	0.53	0.35	0.27	0.17	0.17	0.44	0.54
Cr$_2$O$_3$	0.22	0.22	0.22	0.22	0.33	0.33	0.33	0.22	0.22	0.23
CuO	0.09	0.09	0.09	0.09	0.09	0.09	0.09	0.09	0.09	0.10
Fe$_2$O$_3$	29.73	28.22	27.86	30.07	23.87	22.29	20.88	19.96	27.32	30.26
K$_2$O	0.09	0.09	0.09	0.09	0.18	0.27	0.18	0.18	0.18	0.09
La$_2$O$_3$	0.26	0.18	0.18	0.18	0.18	0.09	0.09	0.09	0.18	0.18
MgO	0.37	0.37	0.37	0.38	0.38	0.25	0.25	0.25	0.25	0.26
MnO	4.73	4.17	4.37	2.64	2.54	2.93	1.64	2.10	1.27	0.90
Na$_2$O	25.08	27.05	27.47	26.68	26.56	26.31	25.62	27.13	24.51	23.42
Nb$_2$O$_5$	2.54	2.68	2.66	2.61	2.67	2.53	2.66	2.83	1.88	1.75
NiO	0.86	0.48	0.77	0.39	0.29	0.39	1.42	1.32	1.15	1.08
PbO	0.40	0.32	0.32	0.24	0.16	0.16	0.16	0.16	0.33	0.33
SiO$_2$	3.43	4.68	5.15	6.74	8.08	7.96	6.55	5.72	3.04	2.31
ThO$_2$	0.43	1.54	2.14	0.60	0.00	0.00	0.00	0.00	0.00	0.00
TiO$_2$	10.67	10.69	10.64	10.91	11.03	10.80	10.79	10.79	10.72	10.04
U$_3$O$_8$	1.41	0.80	1.16	0.54	1.25	2.32	11.97	13.54	12.13	11.87
ZnO	0.00	0.09	0.09	0.19	0.09	0.19	0.19	0.09	0.09	0.10
ZrO$_2$	2.27	2.47	2.46	2.32	2.46	2.36	2.35	2.47	1.78	1.70

Note that the sludge projections did not include sulfate concentrations; therefore, a SO$_4^{2-}$ concentration of 1.0 wt % was assumed for each sludge batch. Noble metals are not typically tracked in sludge batch projections. Therefore, the noble metals Ag, Pd, Rh, and Ru, along with SO$_4^{2-}$, were added to the sludge compositions, followed by a normalization of the remaining components to 100 wt %. The concentrations of the noble metals were obtained from recent measurements of Sludge Batch 6 (on a total solids basis), which was considered to contain a high concentration of noble metals.[11]

A single frit composition was identified that produced a glass composition that was acceptable, based on the DWPF PCCS models, at a targeted waste loading of 40 wt % with each of the sludge composition projections given in Table I. The composition of this frit, which was labeled Frit 0607, is 10 wt % B$_2$O$_3$, 6 wt % Li$_2$O, 5 wt % Na$_2$O, and 79 wt % SiO$_2$. Each of the sludge compositions (in the order shown in Table I) with SO$_4^{2-}$ and noble metal oxides added was then combined with Frit 0607 at a waste loading of 40 wt % to give the targeted glass compositions for the study, which were labeled as the KT08-series.

Each of the study glasses was prepared from the proper proportions of reagent-grade metal oxides, carbonates, and boric acid in 200 g batches. The raw materials were thoroughly mixed and placed into platinum/gold, 250 ml crucibles. The batch was placed into a high-temperature furnace at the melt temperature of 1150 °C. The crucible was removed from the furnace after an isothermal hold for 1 hour. The glass was poured onto a clean, stainless steel plate and allowed to air cool (quench). The glass pour patty was used as a sampling stock for the various property measurements described below.

Approximately 25 g of each glass was heat-treated to simulate cooling along the centerline of a DWPF-type canister[12] to gauge the effects of thermal history on the product performance. This cooling schedule is referred to as the canister centerline cooled (CCC) heat treatment. Visual observations of both quenched and CCC glasses were documented.

Representative samples of each quenched and CCC glass were analyzed by XRD to identify any measureable crystallization. Chemical analysis was performed on a representative sample from each quenched glass to confirm that the as-fabricated glasses met the targeted compositions. Two dissolution techniques, sodium peroxide fusion (PF) and cesium hydroxide fusion (CH), were used to prepare the glass samples, in duplicate, for analysis. Each of the samples was analyzed, twice for each element of interest, by Inductively Coupled Plasma – Atomic Emission Spectroscopy (ICP-AES). Glass standards were also intermittently measured to assess the performance of the ICP-AES instrument over the course of these analyses.

The PCT Method-A[13] was performed in triplicate on each quenched and CCC glass to assess chemical durability. Also included in the experimental test matrix was the Environmental Assessment (EA) benchmark glass,[14] the Approved Reference Material (ARM) glass,[15] and blanks from the sample cleaning batch. Samples were ground, washed, and prepared according to the standard procedure.[13] Fifteen milliliters of Type-I ASTM water were added to 1.5 g of glass in stainless steel vessels. The vessels were closed, sealed, and placed in an oven at 90 ± 2 °C where the samples were maintained at temperature for 7 days. Once cooled, the resulting solutions were sampled (filtered and acidified), then labeled and analyzed by ICP-AES. Samples of a multi-element, standard solution were also included in the analyses as a check on the accuracy of the ICP-AES instruments used for these measurements. Normalized release rates were calculated based on the measured compositions using the average of the common logarithms of the leachate concentrations.

The viscosity of the glasses was measured following Procedure A of the ASTM C 965 standard.[16] Harrop and Orton high temperature rotating spindle viscometers were used with platinum crucibles and spindles. The viscometers were specially designed to operate with small quantities of glass to support measurements of radioactive glasses when necessary.[17,18] A well characterized standard glass was used to determine the appropriate spindle constants.[18,19] Measurements were taken over a range of temperatures from 1050 to 1250 °C in 50 °C intervals. Measurements at 1150 °C were taken at three different times during the procedure to provide an opportunity to identify the effects of any crystallization or volatilization that may have occurred during the test. The data were fit to a Fulcher equation[20,21] to provide a measured viscosity value at the nominal DWPF melt temperature of 1150 °C.

RESULTS AND DISCUSSION

Crystallization
Visual observations of the quenched versions of the KT08-series glasses identified no visible crystallization. All of the quenched glasses were found to be amorphous by XRD. For the CCC versions of the KT08-series glasses, visual observations identified a small amount of surface crystallization on compositions KT08-01, -02, and -03. All of the CCC glasses were found to be amorphous by XRD with the exception of glass KT08-07. This indicates that the volume of surface crystallization in compositions KT08-01, -02, and -03 was below the XRD detection limit. Glass KT08-07 contained a small amount of trevorite, which may have been difficult to identify visually in the bulk of the glass. Spinels, including trevorite, are the crystalline phase typically found in DWPF-type glasses and have been shown to have no practical impact on durability.[22]

Measured Composition
The measured composition data for the study glasses were carefully reviewed. Minor issues with some of the measurements for the glasses prepared by CH were identified. There was likely a minor

batching error for CuO in composition KT08-01. Overall, there were no indications of any significant issues in the batching of the KT08-series glasses. Decisions were made regarding which preparation method would be used for each oxide in determining the average measured composition. These decisions are summarized in Table II.

Table II. Preparation Methods Used in Determining the Concentration of Individual Oxides in the KT08-Series Glasses.

Oxide	Preparation Method(s)	Oxide	Preparation Method(s)
Al_2O_3	CH and PF	MnO	CH and PF
B_2O_3	CH	Na_2O	CH
BaO	CH and PF	Nb_2O_5	PF
CaO	CH	NiO	PF
Ce_2O_3	CH	PbO	CH
Cr_2O_3	CH	$SO_4{}^{2-}$	CH
CuO	CH and PF	SiO_2	PF
Fe_2O_3	CH	ThO_2	CH
K_2O	CH	TiO_2	CH
La_2O_3	CH and PF	U_3O_8	CH
Li_2O	CH and PF	ZnO	CH and PF
MgO	CH and PF	ZrO_2	CH

The data resulting from the preparation methods listed for each oxide in Table II were averaged to determine a representative chemical composition for each glass. A sum of oxides was also computed for each glass based upon the measured values. Glasses KT08-01, -02, and -07 each had one measured value for SiO_2 that was an outlier. These values were omitted as the average SiO_2 concentrations were determined for these glasses. All of the sums of oxides for the KT08 glasses fall within the PCCS acceptability interval of 95 to 105 wt%. A statistical review of the measured versus targeted compositions, which is detailed elsewhere,[23] suggested only minor difficulties in meeting the targeted compositions for the KT08-series glasses, none of which should impact the outcome of the study. The measured composition of each of the KT08-series glasses is reported in Table III.

Table III. Measured Compositions (wt %) of the KT08-series glasses.

Oxide	KT08-01	KT08-02	KT08-03	KT08-04	KT08-05	KT08-06	KT08-07	KT08-08	KT08-09	KT08-10
Al$_2$O$_3$	6.06	5.36	4.66	5.28	7.22	7.59	5.35	4.59	5.21	5.34
B$_2$O$_3$	6.01	5.98	6.01	5.98	6.01	6.01	6.06	5.89	6.02	5.98
BaO	0.07	0.07	0.07	0.07	0.07	0.07	0.07	0.06	0.07	0.07
CaO	0.98	0.98	0.96	0.92	0.98	1.04	0.84	0.74	0.92	0.93
Ce$_2$O$_3$	0.29	0.30	0.27	0.24	0.17	0.13	0.11	0.10	0.20	0.23
Cr$_2$O$_3$	0.09	0.09	0.08	0.09	0.13	0.13	0.13	0.09	0.09	0.09
CuO	0.08	0.05	0.05	0.05	0.05	0.05	0.04	0.05	0.04	0.04
Fe$_2$O$_3$	11.82	11.03	10.85	11.79	9.65	8.94	8.32	7.94	10.73	12.11
K$_2$O	0.21	0.23	0.23	0.22	0.24	0.27	0.27	0.25	0.28	0.22
La$_2$O$_3$	0.09	0.06	0.06	0.07	0.07	0.04	0.04	0.03	0.07	0.07
Li$_2$O	3.53	3.51	3.47	3.44	3.46	3.43	3.55	3.47	3.51	3.45
MgO	0.15	0.14	0.15	0.14	0.15	0.11	0.10	0.09	0.10	0.11
MnO	1.88	1.62	1.74	1.04	1.02	1.17	0.65	0.82	0.50	0.36
Na$_2$O	13.32	13.92	14.19	13.88	13.99	13.75	13.58	14.02	13.01	12.60
Nb$_2$O$_5$	0.86	0.92	0.97	0.92	0.98	0.94	0.89	1.11	0.65	0.64
NiO	0.34	0.20	0.32	0.19	0.14	0.18	0.57	0.51	0.45	0.46
PbO	0.21	0.12	0.12	0.09	0.12	0.12	0.06	0.13	0.13	0.16
SiO$_2$	45.92	48.49	52.09	50.81	51.93	52.52	47.21	48.67	45.67	50.70
SO$_4^{2-}$	0.26	0.26	0.26	0.26	0.26	0.26	0.27	0.27	0.26	0.27
ThO$_2$	0.17	0.62	0.89	0.24	0.02	0.02	0.02	0.02	0.02	0.02
TiO$_2$	4.12	4.16	4.13	4.26	4.35	4.21	4.27	4.18	4.20	3.93
U$_3$O$_8$	0.60	0.44	0.63	0.24	0.50	0.77	3.73	5.22	4.86	4.54
ZnO	0.01	0.04	0.04	0.08	0.04	0.08	0.07	0.04	0.04	0.05
ZrO$_2$	0.82	0.81	0.82	0.74	0.91	0.86	0.77	0.90	0.57	0.62

Durability

One of the quality control checkpoints for the PCT procedure is solution mass loss over the course of the seven day test. Water loss was in the acceptable range for all of the KT08 PCT vessels. One of the vessels, the first replicate of the quenched version of glass KT08-04, had an insufficient amount of glass to meet the required ratio of leachant volume to mass of ground glass. Data for this vessel were omitted from further analyses. This omission does not impact the outcome of the study since each glass was tested in triplicate. All of the measurements of the ARM reference glass fell within the control ranges.[15] A statistical review of the ICP-AES measurements of the PCT leachates, which is detailed elsewhere,[23] suggested only minor scatter in the triplicate values for some analytes for some of the glasses.

The PCT leachate concentrations were normalized using the measured cation compositions of the glasses to obtain g/L leachate concentrations following the ASTM procedure.[13] The resulting values are given in Table IV. The KT08-series glasses all had normalized release for boron (NL [B]) values that were well below the 16.695 g/L value of the benchmark EA glass.[14] The highest NL [B] was for glass KT08-03, with values of 0.65 g/L and 0.60 g/L for the quenched and CCC versions of this glass, respectively.

Table IV. Normalized PCT Results for the KT08-Series Glasses.

Glass ID	Heat Treatment	NL B (g/L)	NL Li (g/L)	NL Na (g/L)	NL Si (g/L)
ARM	ref	0.43	0.51	0.49	0.27
EA	ref	14.42	8.01	11.28	3.64
KT08-01	CCC	0.52	0.61	0.55	0.41
KT08-02	CCC	0.55	0.64	0.60	0.41
KT08-03	CCC	0.60	0.69	0.65	0.41
KT08-04	CCC	0.54	0.66	0.59	0.41
KT08-05	CCC	0.45	0.59	0.52	0.39
KT08-06	CCC	0.45	0.62	0.53	0.39
KT08-07	CCC	0.50	0.63	0.57	0.46
KT08-08	CCC	0.58	0.70	0.64	0.48
KT08-09	CCC	0.56	0.71	0.59	0.48
KT08-10	CCC	0.55	0.68	0.54	0.41
KT08-01	Quenched	0.55	0.68	0.61	0.44
KT08-02	Quenched	0.57	0.67	0.66	0.42
KT08-03	Quenched	0.65	0.73	0.74	0.42
KT08-04	Quenched	0.56	0.68	0.65	0.42
KT08-05	Quenched	0.47	0.61	0.55	0.39
KT08-06	Quenched	0.47	0.64	0.58	0.39
KT08-07	Quenched	0.53	0.67	0.61	0.48
KT08-08	Quenched	0.59	0.72	0.69	0.49
KT08-09	Quenched	0.57	0.71	0.61	0.48
KT08-10	Quenched	0.58	0.72	0.57	0.42

The predictability of the PCT responses was evaluated using the DWPF durability models. The predicted PCT values, determined using the measured compositions of the KT08 glasses, were compared with the normalized PCT responses. Figure 1 provides plots of the DWPF models for B, Li, Na, and Si that relate the logarithm of the normalized PCT value (for each of the four elements of interest) to a linear function of a free energy of hydration term (ΔG_p, or del Gp, in kcal/100 g glass) derived from both of the heat treatments of the KT08-series glasses. Prediction limits at a 95% confidence for an individual PCT result are also plotted along with the linear fit. The EA and ARM results are indicated on these plots as well. As shown in the plots, the measured PCT responses for the KT08-series glasses are well predicted by the DWPF models.

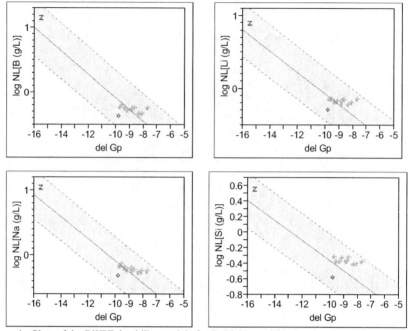

Figure 1. Plots of the DWPF durability models for B, Li, Na, and Si, showing the PCT responses for the study glasses (+), the ARM reference glass (◊), and the EA reference glass (z).

Viscosity

The measured viscosity at 1150 °C was determined by fitting the data for each glass to the Fulcher equation.[20,21] The results of the Fulcher fits were used to calculate a measured viscosity value for each glass at 1150 °C. These values are given in Table V. The measured values are displayed graphically versus the model predicted values in Figure 2. Figure 2 shows that all but one of the KT08-series glasses had measured viscosities that were predictable using the current DWPF viscosity model, based on the measured compositions. Composition KT08-03 had a measured viscosity that fell below the lower confidence interval for the model prediction. However, the difference between the lower confidence interval value and the measured value for this glass is only 2 poise (see Table V), which represents a difference with no practical impact. Overall, the measured viscosity values of the KT08-series glasses are well predicted by the current DWPF viscosity model.

Table V. Predicted and Measured Viscosity Values for the KT08-Series Glasses.

Glass ID	Viscosity Prediction (P)	Lower Confidence Interval for Prediction (P)	Upper Confidence Interval for Prediction (P)	Measured Viscosity (Fulcher Fit at 1150 °C) (P)	Model Predictable
KT08-01	41	28	61	52	Yes
KT08-02	46	31	67	52	Yes
KT08-03	55	38	81	36	No
KT08-04	54	37	78	56	Yes
KT08-05	76	52	111	80	Yes
KT08-06	88	60	129	67	Yes
KT08-07	50	34	74	70	Yes
KT08-08	52	36	76	58	Yes
KT08-09	42	28	61	51	Yes
KT08-10	63	43	93	68	Yes

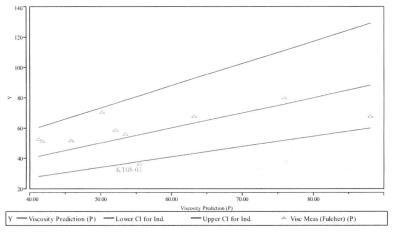

Figure 2. Predictability of the Viscosity Values at 1150 °C for the KT08-Series Glasses.

SUMMARY

A series of glass compositions was selected, fabricated, and characterized to determine the impacts of the addition of CST and MST from the SCIX process on the DWPF glass waste form and the applicability of the DWPF process control models. Specifically, the KT08-series of glasses was designed to evaluate any impacts of the inclusion of uranium and thorium in glasses containing the SCIX components. The glasses were fabricated in the laboratory and characterized using XRD to identify crystallization, ICP-AES to verify chemical compositions, and the PCT to measure durability. The viscosities of the glasses were also measured.

All but one of the KT08-series glasses (both quenched and CCC) were found to be amorphous by XRD. One of the slowly cooled glasses contained a small amount of trevorite, which is typically found in DWPF-type glasses and had no practical impact on the durability of the glass. The measured PCT responses for the KT08-series glasses are well predicted by the current DWPF models. The viscosities of the KT08-series glasses were generally well predicted by the current DWPF model. No unexpected issues were encountered when uranium and thorium were added to the glasses with SCIX components. These results provide confidence that it will be possible to process high level waste with the addition of MST and CST from SCIX at the DWPF. Note however that liquidus temperature measurements remain in progress and may have an impact on the ability to process these feeds. Future work will determine the impact of these compositional changes on the applicability of the current DWPF liquidus temperature model.

ACKNOWLEDGEMENTS

The authors would like to thank Irene Reamer, Phyllis Workman, Pat Simmons, Whitney Riley, David Best, Debbie Marsh, Myra Pettis, David Missimer, Beverly Wall, Mark Jones, Damon Click, and Boyd Wiedenman for their assistance with the fabrication and characterization of the glasses in this study. The authors would also like to thank Dr. David Peeler for his suggestions and helpful discussions. SRNL is operated for the U.S. Department of Energy by Savannah River Nuclear Solutions under contract DE-AC09-08SR22470.

REFERENCES
1. Edwards, T. B., K. G. Brown and R. L. Postles, "SME Acceptability Determination for DWPF Process Control," *U.S. Department of Energy Report WSRC-TR-95-00364, Revision 5,* Washington Savannah River Company, Aiken, SC (2006).
2. Fox, K. M., T. B. Edwards, M. E. Stone and D. C. Koopman, "Paper Study Evaluations of the Introduction of Small Column Ion Exchange (SCIX) Waste Streams to the Defense Waste Processing Facility," *U.S. Department of Energy Report SRNL-STI-2010-00297, Revision 0,* Savannah River National Laboratory, Aiken, SC (2010).
3. Fox, K. M. and T. B. Edwards, "Impacts of Small Column Ion Exchange Streams on DWPF Glass Formulation: KT01, KT02, KT03, and KT04-Series Glass Compositions," *U.S. Department of Energy Report SRNL-STI-2010-00566, Revision 0,* Savannah River National Laboratory, Aiken, SC (2010).
4. Fox, K. M. and T. B. Edwards, "Impacts of Small Column Ion Exchange Streams on DWPF Glass Formulation: KT05 and KT06-Series Glass Compositions," *U.S. Department of Energy Report SRNL-STI-2010-00687, Revision 0,* Savannah River National Laboratory, Aiken, SC (2010).
5. Fox, K. M. and T. B. Edwards, "Impacts of Small Column Ion Exchange Streams on DWPF Glass Formulation: KT07-Series Glass Compositions," *U.S. Department of Energy Report SRNL-STI-2010-00759, Revision 0,* Savannah River National Laboratory, Aiken, SC (2010).
6. Fox, K. M., T. B. Edwards, D. K. Peeler, D. R. Best, I. A. Reamer and R. J. Workman, "High Level Waste (HLW) Sludge Batch 4 (SB4) Variability Study," *U.S. Department of Energy Report WSRC-STI-2006-00204, Revision 0,* Washington Savannah River Company, Aiken, SC (2006).
7. Fox, K. M., T. B. Edwards, D. K. Peeler, D. R. Best, I. A. Reamer and R. J. Workman, "High Level Waste (HLW) Sludge Batch 4 (SB4) with Frit 418: Results of a Phase II Variability Study," *U.S. Department of Energy Report WSRC-STI-2006-00329, Revision 0,* Washington Savannah River Company, Aiken, SC (2006).
8. Raszewski, F. C., T. B. Edwards and D. K. Peeler, "Sludge Batch 5 Variability Study with Frit 418," *U.S. Department of Energy Report SRNS-STI-2008-00065, Revision 0,* Savannah River National Laboratory, Aiken, SC (2008).
9. Johnson, F. C. and T. B. Edwards, "Sludge Batch 6 Glass Variability Study with Frit 418," *U.S. Department of Energy Memorandum SRNL-L3100-2010-00063,* Savannah River National Laboratory, Aiken, SC (2010).
10. Chew, D. P. and B. A. Hamm, "Liquid Waste System Plan," *U.S. Department of Energy Report SRR-LWP-2009-00001, Revision 15,* Savannah River Remediation, Aiken, SC (2010).
11. Lambert, D. P. and A. S. Choi, "DWPF Coal-Carbon Waste Acceptance Criteria Limit Evaluation Based on Experimental Work (Tank 48 Impact Study)," *U.S. Department of Energy Report SRNL-STI-2010-00589, Revision 0,* Savannah River National Laboratory, Aiken, SC (2010).

12. Marra, S. L. and C. M. Jantzen, "Characterization of Projected DWPF Glass Heat Treated to Simulate Canister Centerline Cooling," *U.S. Department of Energy Report WSRC-TR-92-142, Revision 1,* Westinghouse Savannah River Company, Aiken, SC (1993).

13. ASTM, "Standard Test Methods for Determining Chemical Durability of Nuclear Waste Glasses: The Product Consistency Test (PCT)," *ASTM C-1285,* (2002).

14. Jantzen, C. M., N. E. Bibler, D. C. Beam, C. L. Crawford and M. A. Pickett, "Characterization of the Defense Waste Processing Facility (DWPF) Environmental Assessment (EA) Glass Standard Reference Material," *U.S. Department of Energy Report WSRC-TR-92-346, Revision 1,* Westinghouse Savannah River Company, Aiken, SC (1993).

15. Jantzen, C. M., J. B. Picket, K. G. Brown, T. B. Edwards and D. C. Beam, "Process/Product Models for the Defense Waste Processing Facility (DWPF): Part I. Predicting Glass Durability from Composition Using a Thermodynamic Hydration Energy Reaction Model (THERMO)," *U.S. Department of Energy Report WSRC-TR-93-672, Revision 1,* Westinghouse Savannah River Company, Aiken, SC (1995).

16. ASTM, "Standard Practice for Measuring Viscosity of Glass Above the Softening Point," *ASTM C-965,* (2007).

17. Schumacher, R. F. and D. K. Peeler, "Establishment of Harrop, High-Temperature Viscometer," *U.S. Department of Energy Report WSRC-RP-98-00737, Revision 0,* Westinghouse Savannah River Company, Aiken, SC (1998).

18. Schumacher, R. F., R. J. Workman and T. B. Edwards, "Calibration and Measurement of the Viscosity of DWPF Start-Up Glass," *U.S. Department of Energy Report WSRC-RP-2000-00874, Revision 0,* Westinghouse Savannah River Company, Aiken, SC (2001).

19. Crum, J. V., R. L. Russell, M. J. Schweiger, D. E. Smith, J. D. Vienna, T. B. Edwards, C. M. Jantzen, D. K. Peeler, R. F. Schumacher and R. J. Workman, "DWPF Startup Frit Viscosity Measurement Round Robin Results," *Pacific Northwest National Laboratory,* (Unpublished).

20. Fulcher, G. S., "Analysis of Recent Measurements of the Viscosity of Glasses," *Journal of the American Ceramic Society,* **8** [6] 339-355 (1925).

21. Fulcher, G. S., "Analysis of Recent Measurements of the Viscosity of Glasses, II," *Journal of the American Ceramic Society,* **8** [12] 789-794 (1925).

22. Jantzen, C. M. and D. F. Bickford, "Leaching of Devitrified Glass Containing Simulated SRP Nuclear Waste," pp. 135-146 in *Sci. Basis for Nuclear Waste Management, Vol. 8,* J. A. Stone and R. C. Ewing, eds. Materials Research Society, Pittsburgh, PA (1985).

23. Fox, K. M. and T. B. Edwards, "Impacts of Small Column Ion Exchange Streams on DWPF Glass Formulation: KT08, KT09, and KT10-Series Glass Compositions," *U.S. Department of Energy Report SRNL-STI-2011-00178, Revision 0,* Savannah River National Laboratory, Aiken, SC (2011).

Author Index

Audubert, F., 71

Blein, J., 71
Bonnamy, S., 71
Bruneton, E., 71

Cakan, R. D., 1
Carberry, J., 11
Cheeseman, C., 159

David, P., 71
Dominko, R., 1

Edwards, T. B., 169

Feinroth, H., 111
Fox, K. M., 169
Fryxell, G. E., 121
Fukuda, T., 133

Gaberscek, M., 1

Hall, R., 111
Hinoki, T., 25
Holm, E. A., 145
Homer, E. R., 145

Jellison, G. E., 133

Kania, M. J., 33
Katoh, Y., 25, 133
Kondo, A., 133

Markham, G., 111
Matyáš, J., 121
Morcrette, M., 1

Nabielek, H,, 33
Nozawa, T., 95

Ozawa, K., 95

Patel, M. U. M., 1
Pierre, Y., 71

Ramanathan, S., 85
Robinson, M. J., 121

Serre, A., 71
Shimoda, K., 25
Snead, L. L., 25
Sonak, S., 85
Suri, A. K., 85

Takizawa, K., 133
Tanigawa, H., 95
Tarascon, J.-M., 1
Terrani, K. A., 25
Tikare, V., 145

Vandeperre, L. J., 159
Verfondern, K., 33

Zhang, T., 159